Nerds. Geeks. Dorks. Dweebs. You've met us. You know us. You may already be one. We know you'll understand. Welcome to *Noon at the Nerd Table*.

We nerdish learners and knowledge lovers of all ages and genders may often be socially awkward, studious and intense individuals. We pursue our curiosities and passions in depth, sometimes to excess in the eyes of others. And we chuckle at this amazing, always-intriguing world, plus at our own unique, keen obsessions and fascinations to always know more about it.

Enjoy these unique tales and essays as you consider new and interesting angles, ideas, connections, oddities and facts.

And watch out. Nerdism can be fun and contagious. Geek Chic Really Does Rule!

Noon at the Nerd Table

Tales and Essays for the Curious, Passionate and Intense

T.W. King

Copyright 2021 T.W. King
All Rights Reserved.
ISBN: 9798517944481

Noon at the Nerd Table
Tales & Essays for the Curious, Passionate & Intense
First Edition. January 28, 2021
First published by Sunny Cove Publishing PO Box 98 Solon Springs, WI USA 54873-0098.

All Rights Reserved. Except for legitimate use in critical reviews, the reproduction or utilization of this work in whole or in part, in any form by any electronic, mechanical, or other means, now known or hereafter invented, including xerography, photocopying and recording, or in any information storage or retrieval system, is forbidden without written permission of author and publisher: T. W. King, Sunny Cove Publishing, PO Box 98, Solon Springs, WI USA 54873-0098. twkingwrites@gmail.com

Inspired by actual events, this is a work of fiction. Names, characters, places, products, devices and incidents are either the product of the author's imagination or are used fictitiously. Any resemblance to actual persons, living or dead, business establishments, items, goods, products, devices, events or locales is purely coincidental. Essays are factual to the best of the author's current knowledge.

Interior formatting by Hannah Linder Designs

Origins of *Noon at the Nerd Table*

Inspired by actual events, my stories and tales included in *Noon at the Nerd Table* portray fictionalized lives, adventures, events and motivations from diverse perspectives. My essays are factual to the best of my current knowledge. Each piece has been written to highlight the joys and intensity of our collective nerdism. Some stories and essays have appeared in earlier, developmental versions in my other books, periodicals, posts and formats under related or modified titles. - twk -

Various, developmental versions of: Telegraphic Torgeir and Ari, Nerds in Universal Motion, Maddy's Hand-Stitched Towels, Antoinette's Gift, You and Me, Sled Dogs, Spider Friend Wolfie, Debbi's Freaky-Fine Front Flip, Activity Is Best Anesthesia, Grandpa Erny and Pal Peetey, Hard Boiled Mr. Hull, Survival Soup from Scratch, Speaking from the Ocean Floor, Donut Dog Swamps Canoe, True 1890s Farm Story, Power of "On By!", What Party?, Isle Royale Marathon, Full Contact Smelt Run, Treachery and Steel Head Trout, Barrel Stave Skis, Trail Trees Are Forest Magic, Lunch with Colonel Sanders, Park Workers' Disney Spells, Modern Morse Code, Entertaining Garbage, Preschool Drug and Sundries Pusher, Shoes Worth A Fight, Samoyed Raids Valentine Cookies, Tameshiwari: Testing Yourself, Formidable Phillips Code Book, Madeline's Minerals, Y Not UR?, Ice Mirror, Hot Oatmeal, Albedo and Al Gore, Hiking Lake Ice 119 Days, Engineering for the Right Thumb, Dad said Zed. Mom Said Zee, Isle of Pines or Pisle of Ines, Calculus of RLU, Bear Too Hot, Center of the World, Bewildered, Magic Snow Socks, A Sewing Awl, Punkin's Christmas Snacks, Granny Knot or Square Knot?, Children of a Lesser Doc, Winter Love, Treasured Golden Elephant, Just One Day, and Spartacus, Our Durable Squirrel have appeared under various titles in one or more of these publications: Superior Telegram newspaper, QST magazine of ARRL, Dots and Dashes journal of the International Morse Telegraph Club, Wisconsin Writers Association *Creative Wisconsin*, North Star National Park Service magazine of the North Country National Scenic Trail, ISBONA journal of Icelandic Sheep Breeders of North America, Wisconsin Brain Injury Association-Barron County Chapter newsletter, as well as other periodicals, and also in these books authored by Thomas Wayne King: *Tales from the Red Pump* ISBN 978-0-578-04214-5, *Neighborhood of Bears* ISBN 978-1-365-99556-9, *Magic Snow Socks...tales of practical transcendence,* ISBN 978-1-67802-766-7, and *Red Pump Chronicles* ISBN 978-1-716-50532-4. Others not listed here may also be included.

The longest story, "Spartacus, Our Durable Squirrel" appears in Thomas Wayne King's book **Magic Snow Socks...tales of practical transcendence,** 2nd edition, Volume 4 in his **Tales from the Red Pump** book series of Northland stories and history, ISBN 978-1-67802-766-7, Pages 139-148. Sparty's story is also part of King's 2020, 700-page compilation of the best from all four of his volumes, **Red Pump Chronicles**, ISBN 978-1-716-50532-4, pages 651-657. An abbreviated, developmental version of his Spartacus story is published in **QST**, international journal of the American Radio Relay League (ARRL.org), Newington, CT. This true Northland squirrel tale also appears in the St. Croix Writers 2020 Anthology, **Many Waters**, ISBN 979-8-68912-674-6, ages 58-64.

Please Note: While this book is intended for the enjoyment of general audiences, a few portions of *Noon at the Nerd Table* contain words, terms and concepts that some readers may find disturbing or offensive.

For great grandparents, Thomas Favell and Angela Haste Favell, both writers of great skill, who inspire, teach and motivate me from more than century and a half after their first works appeared. Thanks Tom and Angela. So glad to get to know you through your wonderful words you left to us.

For Rat and Mac, everyone at Kronshage Hall, and Puritans of '68. Wherever you guys are now in the Geekiverse, here's to you.

For Leon, first lizard. Chameleonic charmer, muse, mentor, inspirer.

For all my loquacious, learned colleagues of St. Croix Writers Group, who have heard, read and critiqued my work for so many years. You keep me going. Thanks always.

For Debbi, my best friend, lover, wife, first editor, encourager and subject of the best stories I will ever write. Let's keep on living them, every day, throughout every new, fun, odd year together. May curious, passionate and intense always define us.

Contents

Preface	xv

PART I
PEOPLE & CRITTERS — 1

1. Good Morning, Wolfie!	3
2. Tom's Digger World	6
3. Lunch With A Stunning Prisoner	9
4. Telegraphic Torgeir And Ari	12
5. A Difficult Decision	14
6. Freshwater Mussels	17
7. Maddy's Handstitched Towels	20
8. Antoinette's Gift	23
9. You And Me And Sled Dogs	26
10. Noon At The Nerd Table	29
11. Debbi's Freaky-Fine, Front Flip	32
12. Activity Is Best Anesthesia	34
13. Grandpa Erny And Pal Peetey	37
14. Hardboiled Mr. Hull	40
15. Survival Soup From Scratch	43
16. Donut Dog Dunks Canoe	47
17. TRUE 1890s FARM STORY	50
18. Power Of On By!	53
19. What Party?	55
20. Isle Royale Marathon	58
21. Full Contact Smelt Run	62
22. Kruska Lunch In Reykjavik	65
23. Samoyed Raids Valentine Cookies	68
24. Treachery And Steel Head Trout	71
25. Simple Barrel Stave Skis	73
26. Lunch With Colonel Sanders	77
27. Park Workers' Disney Spells	79

28. Hiking Lake Ice 119 Days   82
29. Punkin's Christmas Snacks   84
30. Preschool Drug And Sundries Pusher   87
31. Shoes Worth A Fight   90
32. Treasured Golden Elephant   92
33. Just One Day   95
34. Magic Snow Socks   98
35. Spartacus, Our Durable Squirrel   103

PART II
SCIENCE & TECH   111

36. Nerds In Universal Motion   113
37. Speaking From The Ocean Floor   115
38. Trail Trees Are Forest Magic   119
39. Tameshiwari: Testing Yourself   121
40. Formidable Phillips Code Book   123
41. Modern Morse Code In Rehabilitation And   126
    Education
42. Madeline's Minerals   129
43. Hot Oatmeal, Albedo, And Al Gore   132
44. Engineering For The Right Thumb   135
45. Dad Said Zed. Mom Said Zee   137
46. Isle Of Pines Or Pisle Of Ines   140
47. Calculus Of Rlu   143
48. Bear Too Hot   146
49. Center Of The World   148
50. A Sewing Awl   151
51. Granny Knot Or Square Knot?   154
52. Cleopatra: Word Nerd   156
53. Children Of A Lesser Doc   159
54. Winter Love   163
55. Conspire   165
56. Eternity With Pinewood Mercury?   168
57. Ice Mirror   171
58. Y Not Ur?   175
59. Stereognosis   178
60. Perseverance And Ingenuity   181

61. When Nerds Post Noontime, Office Door Notices	184
62. Entertaining Garbage	186
Thank you	189
About the Author	191

Preface
Welcome To Nerdom

Nerds, geeks, dorks and dweebs abound. We are everywhere, in all walks of life, even here, in our rustic, old-growth-forest village in northwestern Wisconsin. These sixty divergent, tales, stories and essays of applied, personal geekitude illustrate and celebrate our many sides of NGDD life. Revel with those of us who live to learn, study, analyze and experience life in all its flavors and nuances. We are curious and passionate, pursuing details, depth and expertise with intensity. Sometimes a bit too much. But, oh, do we enjoy it.

Nerd life can have its rough patches. Oh yeah. Yet, we can do well, succeed abundantly, and even win it all. You are invited to share these dweebish times and adventures. Your noon seat at our separate table is waiting. You know you belong. Bring your favorite protractor, cartridge pen and mechanical pencil. Be ready to laugh, weep and cringe at our best and worst.

This collection of sixty-two oddly cool chapters includes several of my earlier works, enhanced now from my growing perspective of pervasive, urban and rural nerdism. You will find it apparent that situations I write about exemplify true geekish focus, inquisitiveness, and pursuit of things arcane and difficult: titles and topics definitive of active, studious, passionate lives.

Tales and essays here are grouped into two loosely determined categories: *People & Critters*, and *Science & Tech*. Sometimes our NGDD factor is evident in the focus and adventures of the characters, with both human and animal actors. Sometimes, the essays and tales simply nerd-out on interesting topics that beckon, deserving more exploration. Hope you will check into them. The groupings are arbitrary and interchangeable. You may interpret them any way you want.

So, allow your fascinated, passionate, curious self to enjoy these wordy delights. Nerds, geeks, dorks and dweebs of all types and species are everywhere. We can, will and often do rock the world...if we can just find our glasses, phone and keys. *Oh well...*

T. W. King January 28, 2021 Solon Springs, WI USA

Part One
PEOPLE & CRITTERS

Chapter 1

Good Morning, Wolfie!
A Man and His Spider

"Good morning, Wolfie! Did you miss me?" My workday in our brand-new barn attic began with sorting, stacking, recycling, and discarding, all such important parts of moving.

Wolfie, a formidable, mouse-size, dark-brown Wolf spider, had taken up residence in the attic of our just-built, old-growth-forest barn, living and lurking high up in the corner of two angled roof joists. Wolfie established a daunting presence there. I confess that I needed to get used to *her*. Yes, her. She was a large, impressive Momma Wolf Spider.

Unique among arachnids, Wolf spiders are of the family *Lycosidae,* from Greek for wolf, who don't spin webs. They prowl and predate directly, using stealth and their fangs to attack and gobble their prey of spiders, insects, and even small vertebrates. But not me. Scared? Well...

Wolfie was big, but not that big. Even so, any other bugs and spiders in our shared attic world were her fair game. She was the one and only noticeable spider living up there as I worked to move in items from our previous homes and garages during our intense three-year period of construction, and of persistent lugging of bulky things upstairs.

Wolfie became attic active as we moved in our stored boxes and furniture from residences in southern Wisconsin. These boxes, couches, beds, shelves and chairs had spider webs on and in them from years of storage, housing many smaller spiders of various types that could include Brown Recluse and Black Widow, among other more common and numerous house spiders that live mostly unnoticed so close to us. I'd see them skitter occasionally as we moved load after load of our stuff from home to trucks, trucks to barn, and then up the new attic steps over any months. Spiders, alive and dead, were along for the ride, in considerable abundance. We kept Wolfie well fed in those early days, and she served admirably as fierce terminator of unwanted crawlies. She was top attic predator: our resident arachnid T-Rex.

We gradually got all our stuff into our new barn attic, and it has settled in there over the several years since. I still saw Wolfie into the sixth and seventh years of our barn, in her corner, high up in the concealed nook of conjoined roof beams. I haven't spotted her or family since, but there seem to be no other spiders or bugs in the attic now. No webs, no eggs, no dead or live spiders; no flies or critters at all that I can find. So maybe it still is the Wolfie effect, with her staying out of sight now that we have more items stored up there to conceal her. And kids?

Over those first years, Wolfie was a plump, two-gram brown dot on the ceiling beam who watched me closely each time I came into her territory. She would stay out in the room, high up on the joists observing me, angling her body and backing up a bit sometimes, but always staying firmly in her lofty corner, always keeping her eight eyes on me and potential meals I may bring along in new boxes of stuff I carried in.

Although it has been a while, I believe Wolfie continues to be our busy, effective cleaner. It was and still is tough to be a visiting spider or bug in our barn attic. None are there. The big, bad Wolf spider apparently gets them all when they move in. Our Wolfie, with her formidable fangs and ability to silently stalk, chomps and eats them soon after their arrival. I am glad she is my silent friend...and that, at 185 pounds, I am more than ninety thousand times heftier than she is. Even though I have no fangs, she and I both know full well who can squash whom in our on-the-job relationship.

"So... Where are you, girl?" I have been thinking and asking this aloud when I work now to further straighten and clean up our attic storage space. I miss her. Perhaps, in his next western movie, Kevin Costner might have a role for me as "Talks with Spiders."

Nowadays, years after erecting our new barn, I keep watching for Wolfie and her many-times-over descendants. I believe they keep on policing up undesirable creepers, possibly even a few poisonous spiders that traveled along into our storage spaces. We have an unspoken deal: she doesn't bother me, and I am careful where I set new, spider-crunching boxes. It seems to be working well. Our barn attic remains clean, almost unnaturally pristine, although I haven't seen Wolfie and her kin for some time. Maybe she still lurks in the corners and shadows?

Just to be positive and polite on this new day, I will soon go up into the barn attic to work, and will call out to our silent, sentient sentinel "Good Morning, Wolfie! Did you miss me? Hope you are still here."

Chapter 2

Tom's Digger World

Recently, rumors circulated through our forested neighborhood, here in our quiet little village in far northwestern Wisconsin. Lunch with our neighbors, along with a few beers, got us agitated about the possible approval by our village board of a seasonal trailer and RV park just down our peaceful streets. One of our other neighbors was exploring ways to use his forested acreage to generate income on his land. He would soon bring forward his plan of clearing his forest and installing infrastructure for a multiple-unit recreational vehicle park near all of us.

We close-by families were horrified at what could happen if we did not act. It all seemed so silly and disruptive in our shared, peaceful woodland paradise. So, not to be outdone, and to stimulate discussion about the inappropriateness of his wacky proposal, apparently supported by his friends on the deciding committees, I prepared my even-sillier proposal to put forward to the village board, regarding *my* family's plans for *our* five-acre property just down the road.

Announcing:
TOM'S DIGGER WORLD, DRIVE-IN THEATER, and RV* PARK!

****Reste Verde***: Green Rest Natural Burial.
Don't delay. Play. Stay. Someday Decay!

Local, rural resident and fifth generation landowner, Tom King, unveiled today his plans for a unique, new digger park, drive-in theater complex with rock bands and fireworks, and the region's only natural burial center: ***Reste Verde***, meaning Green Rest, to be located just east of the local funeral home, across the business highway.

Reste Verde will also have express-service at the village sewer ponds, where, if it is your final wish, you can be instantly interred.

"We need to attract more commerce to our peaceful little village in the forest," Tom said. "So, let's remove these annoying huge trees on our five-acre parcel, pave over half of it, and bring in backhoes and bull-dozers on the rest. Guys can just go at it. Dig and have fun. No rules!"

King encourages everyone who has always wanted to excavate dirt, and move sand and gravel around with power equipment, to stop by and watch his progress in building this much-needed complex in their pristine residential neighborhood. He notes that their natural-burial option allows you to have a purpose for your recreational digging.

Bring a friend. Living or not.

"Come for the movie, enjoy the machinery, and make this your place of eternal rest when the time arrives," he stated. "***Reste Verde*** is the best full-service RV Park you will find in this region. You will love our drive-in movies on our newly paved, bug-free acreage. Just think, it was previously being wasted on useless, leaf-dropping trees, and all those annoying, random, pooping wild animals and birds. Now it will be valuable!"

King is certain his neighbors will delight in these planned improvements. He knows they will stop by, eager to operate diggers, watch films, and visit friends and loved ones when eternity beckons.

"True Progress marches on in our idyllic village!" King stated, beaming with rural pride.

Chapter 3

Lunch With A Stunning Prisoner

"May I join you?" Surprised and speechless, I nodded agreement to her as I ate my bag lunch, motioning to a clear area on the lawn near me.

On that day, I was an undergrad human lab rat, employed in the nuclear medicine department of UW Hospitals in Madison, Wisconsin. Our team started early each day, often working through our morning breaks, due to complex quirks with our expensive research equipment. Our half-hour noon lunch times were treasured by our group. The real, doctoral researchers, along with we lowly student summer employees, needed a calming intermission in our intense data gathering. Significant, daunting levels of radiation were always factors in our work.

To escape our shielded laboratory confines, I carted my bag lunch outside each clear day, dining as I sat on the expansive front lawn of our sprawling hospital and clinic complex. Typically, I gobbled my simple self-packed sandwich, cookies and apple, then reclined onto soft grass, in the shade of low cedar trees, resting my eyes and brain until I heard the university carillon chime on the half hour.

Other workers, along with students from all divisions of our broad medical community, bustled around me. People were in white and blue coats, suits, dresses, scrubs of varying colors, shorts, t-shirts, jeans...the whole gamut of clinical and student summer clothing. A mellow crowd, we all relaxed our own way at noon. Sometimes we'd detect a distinct, burning-vegetation odor emanating from secluded, shaded groves of high bushes nearer the buildings.

Usually, I was with one or two grad students or undergrads, and faculty and clinicians often joined us. But some days, I sought out my own secluded spot in the shade, and plunked my tired self down.

This day, I had been eating and dozing. But not now. She stepped up close to me in her staff uniform, and sat down with her lunch. Stunning, with brilliant blue eyes and waist-length strawberry-blonde hair, she sat on the grass next to me, her lunch bag and carton of milk in hand.

Petite and fit, she spoke softly. "Hi!" She told me her name, asking "What department are you in?"

"Uh...Hi," I blurted. "Tom." My blundered, nervous reply caused me to miss her name. Rattled, I forgot to say my lab office and job, and I was too flustered to look at her hospital name badge.

Dressed in the light-blue garb of housekeeping staff, she described her job of cleaning patient rooms and offices in our huge complex. I noticed another ID tag on a bright lanyard around her neck. *County Jail...* something. We talked about the fine day and got acquainted.

Opening her lunch bag, she sighed, "Wonder what my prison sandwich is today?"

"Prison?" I interrupted.

"Yes. Shop lifting. Thirty days. They got me. Little silk scarf in a store. Thought no one noticed. Don't know what I was thinking."

She swept back her flowing red-golden hair, enjoying the sun, as we shared my homemade chocolate chip cookies. She loved them. I got up to return to work. She touched my arm, inquiring with her intense bright-blue eyes. "Will I see you tomorrow?"

Chapter 4

Telegraphic Torgeir And Ari
Our Morse Code Sheep

Torgeir, coal black, and his cousin, Ari, cloud white, are wise, human-experienced characters in our flock of tenacious, all-weather registered Icelandic sheep. Both are senior, nine-year-old spokes-sheep for our group of rugged, woolly individuals. And each knows how to read their shepherds, Debbi, N9GLG, and me, WF9I, both amateur radio operators. At our primitive, northern Sunny Cove Farm, they have figured us out.

Each day, we feed our flock hay, renew drinking water, clean pens and sheds, and complete chores needed to keep our treasured small ruminants healthy and secure. Their luxurious fleece and amazing 1,200-year-old eyes from Viking times, encourage us to do our best for our durable charmers. They look forward to their mid-morning brunch.

As we go about daily chores, we rotate sheep among five, fenced grazing areas to reduce their exposure to parasites present in soils and grasses, making certain each paddock is well-protected from predators. Here, in our far-northwestern area of Wisconsin, overlooking bays and harbors of Lake Superior, our county has twenty-eight packs of wolves. We also have bears, coyotes, foxes, fishers, and an occasional cougar as forest neighbors. And we get along fine. Over our twelve years of raising valuable sheep in our remote location, we've had no issues with predators. Careful, wise engineering of fences and sheds has worked well, and our tenacious sheep are smart, alert dwellers within our surrounding expanse of wild forest. Their paddocks are their whole world: their safe, protective sanctuary.

As we finish hours of farm chores each day, we look out over our interconnected grazing paddocks, where the sheep have spread out into their secure areas to feast on grass and wild treats. In her clear singing voice, Debbi calls them back to their main paddock with her grain song. "It's grain. It's grain! It's oats. It's corn. It's treats..." Her song works. They stampede toward her as she hurries scoops of enticing grain-treat mixture to their main feeder. At the first words of *grain* or *treats,* they alert, charging like a football-team front line, almost running her over.

Dangerous stuff. We've tried spelling out the words or just saying first letters, but the flock figured that out. Aha! Morse Code to the rescue. We deployed parental Morse "CW" code signals with our boys when they were little so they wouldn't know our words. It worked, but then they also became FCC-licensed amateur radio operators (Hams) at ages nine and ten, and knew exactly what we were up to.

With our sheep, we also attempted spoken, first-letter CW. "Dah Dah Dit" for *G*. "Dah" for *T*. Out of all our flock members, Torgeir and Ari caught on. What now? Traffic signs, semaphore, finger spelling, sign language, whistling...? SOS!

Chapter 5

A Difficult Decision

A brilliant young woman, "Helen" was nineteen years old when she died in 1886. Her wealthy lumber-baron family built an ornate library and theater complex to honor her memory. It still occupies a revered position in the western Wisconsin town where I grew up. My parents, schoolmates and I attended and participated in lectures, concerts and plays there for years. I read most of their science fiction and paleontology books in my K-12 years, 1955 through 1968.

During my high school days in the little town, and since, the cause of Helen's death has been a topic for continuing conjecture and discussion at lunch times, and after hours in bars. The official record conveys that "cancer of the side" (appendicitis?) was the cause. That may be correct.

But it might be something else. Accidental, complicated and unwanted pregnancies have always presented challenges for women and men, families, and society in general. How does a culture regard and care for single mothers, plus the fathers, and all of their families?

For Helen, could it, in fact, have been a failed abortion that ended her exceptional life so early? We may never know with certainty in her case, but more than half a century later, during the 1960s, vivid discussions about rumored abortions available in that area were still common.

And they included frequent mentions of "Coat Hanger Fran." Was she fact, legend or something else? Some were assured of her existence.

Near the local university is an older brick building of 1920s vintage that has housed various offices and stores over the last century. It is well known, and frequently visited for its various retail functions. Not fully obvious to most passersby is a side railing, and a shadowed stairway disappearing down to the basement level, along the secluded side wall of the building. The steps and railing are hidden in darkness, well-shaded, and tucked among close, neighboring structures.

A half century ago, regional tales held that Coat Hanger Fran practiced in that basement office. Her hidden clinic served desperate women in confidence: women who were likely alone, poor and frightened. I've heard this, but have never confirmed her existence. Fran was spoken of by reliable friends, both women and men. Her services were said to be available to anyone who paid cash. Area student populations of the time were alert to her presence, her passion, and her fees if they needed her.

The primitive method used then was likely an extended coat hanger, or other thin wire, inserted through the woman's cervix, then turned and twisted to dislodge anchoring fetal cells in her uterus. Such homespun surgeries were both unsanitary and dangerous. Even if deployed with the best of helping intentions, the brutal procedure could not override its inherent lethal imprecision. As sharp, ragged wire ends snagged and tore delicate tissues, possibly ripping into branches of the uterine artery, the subject could die of resulting infection and blood loss.

If Coat Hanger Fran and her mission were real, how many

terrified women, often alone, desperate and impoverished, lost health and life because they neither knew of nor could embrace other viable options?

Chapter 6

Freshwater Mussels

THIS PART IS TRUE:

Along the upper Mississippi River, in large metropolitan areas, purification of river water for drinking has a unique, initial step, carried out by unusual city workers. It first gets sampled by sensitive sentinels in the form of durable, hard-shelled filter-feeders: freshwater mussels. *Mucket Clams*, also known as *Actinonaias Ligamentina*, are kept in containers and exposed to the flow of new water entering urban purification systems. The alert mussels clamp their shells shut if they sense even minor impurities in their surrounding waters. Water-quality staff explain that, just as humans balk when they smell something threatening around them, these clams close up tight when they sense a chemical threat in their liquid environment. Receiving this important warning signal from Mucket Clams can alert metropolitan water supply workers to dangerous, noxious impurities that may otherwise enter water-storage tanks unnoticed. Several U.S. towns and cities rely on trustworthy clam guardians in their initial water treatment processes.

THIS PART IS NOT TRUE. BUT IT IS FUNNY. ADAPTED FROM WASHROOM-WALL HAND SCRAWLINGS AT A UNIVERSITY:

Unknown to most, Mucket Clams are also excellent for eating. As a gourmet treat, they said to have been bred and refined for decades by the Schmitt family, rural dwellers along the big river. These hand-raised, sensitive and delicious freshwater mussels are known in the food world simply as "Schmitt," both the singular and plural term for the famous family clams. Schmitt are legendary for the difficulty of opening their tight shells before cooking these much-sought mussel delicacies.

Correct shelling of tenacious, succulent clams takes knowledge and skill: an art passed on, person-to-person, by clam lovers preparing freshwater mussels for their table. They are dedicated Schmitt eaters who love big plates of Schmitt, plus Schmitt sandwiches, and Schmitt chowder. Enthusiastic fans are merrily full of Schmitt. They revel at festive, monthly clam celebrations known as Schmitt Storms. If you're interested, you can learn to shell your own Schmitt from expert, volunteer Schmitt shellers.

As you dip piles of Schmitt with your net from along the riverbanks, you will find informal enclaves of eager Schmitt shellers engaged in their arduous work on convenient picnic tables. Want to become a skilled Schmitt sheller, too? This is your opportunity.

"No Schmitt." You might say, hesitating to try something new.

"No problem!" They'll answer. "We have blessed large Schmitt loads; they are waiting for you. Holy Schmitt."

Schmitt shellers are helpful. Bring a big bag of your own fresh Schmitt. They'll gladly shell your Schmitt. They love shelling Schmitt, yours or theirs, and are ready to welcome you to their world of Schmitt.

Don't worry about piles of lingering Schmitt shells. Each Schmitt sheller handles their own dump.

During their active spring and summer Schmitt-shelling seasons, the dedicated group sponsors creative egg-coloring and multi-hued T-shirt dipping events at fresh-Schmitt picnics. Posters on bulletin boards all along the river express their wild enthusiasm: "Eat Schmitt and Dye!"

Chapter 7

Maddy's Handstitched Towels

Maddy, loved to learn, explore, and read about everything. Always engaged with details of the world, Mom spent many of her later years alone at home and in assisted living after her husband of 65 years, my dad, who died at 91. Maddy lived to age 94 and kept busy, occupied with news, politics, sports, gardening and handwork. As her eyesight and hands weakened, she did edging and embroidery on simple, plain, white cotton gauze dish towels, alone, in her quiet home, and eventually in her care center room. We have kept some of those "Maddy Towels" as reminders of Mom's fortitude, and her firm determination to not give in to isolation and depression.

Throughout her long life, Maddy was a jolly, resolute soul who taught us much about living in desperate, hard situations. She survived the long-term suffering of family illnesses, disabilities, and early deaths. She also adapted to and mitigated myriad family challenges and disappointments, as best she could, over the long years. Maddy lived always forward to prosper and triumph over adversities by savoring and enjoying her life each day. She neither gave up on, nor gave in to, a nearly ten-decade lifetime of facing insoluble problems.

Memories of Maddy in her tough times inspire me to carry on and keep going in the face of daunting conditions over which I have no power, no control. But she taught me that I do have choice. I don't have to be overcome by the life obstacles that might attempt to diminish me, and I can fend off fears and anxieties by engaging with projects and people.

When Debbi and I moved to our new home and village soon after Maddy's passing, we packed everything in haste and confusion. Over the years, we have unpacked most of our things.

One of the heart-touching items I found recently is Mom's bag of dish towels that, with great care, she individually crafted, as she stitched along and around their edges, and embroidered in their centers. Each towel is ringed with colorful stitching on all the edges, and, on many, she created designs of flowers and birds stitched into the center areas. We must have twenty or more of these unique treasures: our Maddy Towels. They are of little value as dish towels now, so thin and worn, but they are of immeasurable value to me and those who wish to remember or come to know the spirit and heart of Maddy.

Even in Mom's loneliest, darkest days when Victor, and their daughter, my sister, Karen, were gone, she kept on creating. Maddy created usable personal art to keep her mind and hands busy. And, as I realize now, years later, to give us something practical and tangible to keep, to help us remember her tenacious, triumphant life.

Maddy's dish towels are examples of her many creative, unique ideas throughout her exceptional life. *Maddy Towels* are another of her tangible, poignant solutions to sadness, providing touches of immortality and remembrance. As we continue to learn from Maddy's handstitched towels, and her rugged persistence throughout times, great and trying, we treasure her practical ways of persevering, along with her wise, workable, creative problem-solving strategies that she passed on to us.

Chapter 8

Antoinette's Gift

"Go ahead, Toni. Tighten up those axle straps." Grandfather Roy encouraged, as Antoinette helped him reattach wheels to their Red River cart, wrapping buffalo-hide fasteners with speed and skill. In 1867, their alacrity derived from Roy's decades of fabricating durable Metis carts. Toni, age 13, was now his cart-making partner, as his son, Emil, Toni's father, hunted scarce buffalo for winter.

In eastern Manitoba, the family spoke Michif, their language of Metis people, combining Plains Cree and Canadian French. Durable, adaptable, two-wheel Metis carts, constructed entirely of wooden parts fastened with hide lashings, could go anywhere. Strong and light, the wagons were indigenous genius: quick to take apart to pack on your horse or your back, or to float across a river. Buffalo-hide casings for the huge wooden wheels formed flotation rings to be stacked together as a raft. Other cart sections were lashed on top so that all components could be pushed across the water, then reassembled on the other side.

Eager and fascinated, Toni learned well.

Grandfather Roy liked working with Toni. She was quick,

smart and never afraid to get dirty. He gave her knowledge and skills to craft and assemble each piece, and to take all apart and reassemble them into a rolling wagon, ready for work in all seasons, on all types of terrain.

Oxen or horses could pull, too, and Toni mastered those detailed hitching skills. But she and Roy usually powered and carried their own carts themselves, often through rushing river waters and deep-mud trails.

They were a poor family, so other members walked. Only small children or elderly people rode.

With her father off to hunt buffalo and gather winter foods for their daily meals, and Mother with four little ones to tend and cook for, Toni did what she could to help with sewing and cooking at home.

But Antoinette's full interests and skills were with the carts.

In summer of 1869, they moved southwest of Winnipeg, nearer to the U.S. border. Better growing conditions attracted them, along with the thriving Red River Metis culture. Soon, Antoinette was invited to a local sewing circle by affluent city girls, formerly from Montreal, Gatineau, Toronto, and Ottawa. Grandfather encouraged her to listen and speak in English, as her world of new language opened to Toni.

On a wet, muddy day, three wealthy sisters rode in an older rickety Metis cart to their Holiday sewing-group party. Almost to the meeting house, their wagon broke. In high-button shoes, with frilly dresses and ornate sewing boxes, the girls would have to slog through slush and mud to the celebration.

Toni had just hiked in to join, and rushed to help. She tossed their warming blanket down on the muck, then helped the girls step out. An axle support had broken, but a spare was lashed underneath. Untying hide strappings from near the left wheel and axle, she ducked under and replaced the part. Within minutes, Toni had it fixed. She calmed and re-hitched the horse, motioning

"Get in girls," then led horse and cart to the wide front doors and hitching rail to unload.

One of the haughty, rich young women showed Toni her hefty, gilded packing trunk and expensive sewing baskets. "These are my wonderful gifts from Grandfather in London. Did your grandfather give *you* any special gifts before you came to the Red River Valley?"

Antoinette understood. Pleased, she paused, smiling and gazing at the cart she had just repaired so fast and easily, and at the now-willing, calm horse she secured with compassion and skill. Confidence and satisfaction gained from her young lifetime of experiences with Grandfather Roy rose in her heart. With pride gleaming in her eyes, she nodded and looked at the girl.

Relying for the first important time on her new skills in English, Toni replied, "Yes. Yes... *He did!*"

Chapter 9

You And Me And Sled Dogs

Sled dogs are admirable, determined heroes, amazing in their diverse adaptability, power and tenacity. They run hard and they rest hard, as do the people who love them. A way to think of sled dog types is to group them into one of three categories: racing, cargo, and expedition.

Racing Dogs tend to be fast, small and lighter in weight than the other types. They are selected and trained for running with exceptional speed while hauling smaller loads for sprints or for fast, long-distance runs. These mixed-breed dogs are often referred to as *Alaskan Huskies*, and may include diverse breeds such as Greyhound, Malamute, Siberian Husky, Border Collie, Poodle and others purposely interbred and developed for a fleet-footed mix of intelligence, speed, endurance, and cooperativeness with their team members and drivers.

Cargo Dogs are generally larger in size and weight, with strength and determination for pulling heavy loads at slower speeds through rugged, challenging country. Large, muscular cargo dogs can be of many types and dispositions. They may be of Inuit, Newfoundland, Malamute or other tough, bigger, reliable breeds who can power through and beyond deep snow and steep inclines. Dependable moving of heavy loads over long distances and difficult terrain is essential for them.

Expedition Dogs combine dependable, mid-range characteristics of both racing and cargo dogs. Th3ey are known for their focused stamina over demanding, long-distance hauls. Expedition dogs are not just racers or heavy-load haulers. Rather, they are tenacious, select pullers of the various breeds already mentioned above who combine abilities. They can work for extended periods of exertion, and get along in teams of dogs and humans through challenging, lengthy stretches of trail. When long-distance sled teams have people, supplies, equipment and other essential items to transport, expedition dogs, with the best qualities of the other types, are the determined, persistent work crew who will deliver their loads through rugged terrain and weather.

SINCE BECOMING acquainted with sled dogs decades ago, and then raising and training two dogs to do simple pulling tasks at our primitive farm, I have come to think of others and myself in terms of sled-dogs. Despite its shortcomings, my sled-dog shorthand method summarizes human characteristics and personalities well. Each category of sled dog is exemplary, with unique powers: each a canine superhero.

My personal preference and self-image is expedition dog. I am not the fastest or strongest, although I have some capacities for speed and strength. My talents and nature are more like the tenacious, persistent expedition canines. I am not the quickest, smartest or most-powerful team member, but I am determined to be among the most reliable. I will complete my trek when setting out upon a challenging course of real life, knowing the race in daily human existence goes not so often to the swift, but more often to the gritty and persistent.

Expedition dogs keep on running and hauling on their trails. They may not get there first or with the heaviest loads, but are resourceful, reliable, and will arrive safe and alive. Expedition dogs keep moving forward even in the face of daunting adversities. My canine heroes, all.

So, which kind of sled dog are you? There is no wrong answer.

Chapter 10

Noon At The Nerd Table

High school lunch break. 1967. Brown Baggers. Science and math types. Sophomores through seniors. Twenty minutes. All guys who packed our own lunches, rushing to sit down and eat.

We were a miserly, discerning bunch, giving up on school hot lunch long ago, after their supposed "chipped beef" still had hair on the unchewable, indeterminate meat strips, sogging in curdled, nauseating white sauce. Our trust in hot lunch was forevermore ruined.

Benjamin plopped down. "New song on the radio: *Asshole* Man," he blurted to our group, as he dropped his slide rule and stack of books next to his lunch bag. Mechanical pencils and fountain pens in his shirt-pocket protector wobbled and almost fell out. "*Asshole* Man. Yeah."

Rat glared at him. "What are you talking about?" Rat's real name was also Benjamin, but we called him Ben, or mostly "Rat" because he looked like, well...a rat. Short wiry guy, big nose, huge front teeth, wide ears, small tight eyes, bountiful acne. Yup. "Rat" was what you'd think even if you'd just met him. He liked to be in charge of our misfit bunch.

"You're nuts," Rat muttered. He bit into his peanut butter sandwich and sipped chocolate milk.

We noon-hour nerds at the lunch table looked up from our brown bags. "Huh, *Asshole* Man?" someone queried. "What kind of song is that? What are you listening to?"

"I don't know..." Benjamin shrugged. "Just heard it this morning when I drove in." A brilliant calculus and physics student, he was a farm kid, who drove to school each day from his family place about ten miles out in the country. He had time to listen to the radio in the barn, early, and his ideas were always worth respecting. But he didn't get out into the larger world very often.

As a dual-purpose, science *and* music geek at the table, I responded. "Think I know the song, Benjamin. 'I'm A *Soul* Man' ...by Sam & Dave. S-O-U-L Man. A rhythm and blues, up-tempo thing, with horns, right? Heard it when I was painting yellow street lines for my city job."

"Yeah, you know it!" Benjamin was stunned. "Glad you've heard it, too."

"They play it a lot on KDWB and WDGY." I was always listening to Twin Cities rock, and to upbeat R & B stations in those days. "A Soul Man. Pretty cool song."

"Uh Huh. You got it," Benjamin nodded, as we all finished our lunches and stood up to leave.

"Hey, you losers..." Rat took over. "How does a mathematician solve constipation?" Silence from all. "He works it out with a pencil," Rat guffawed, spewing brown milk, peanut butter and rye bread in full process from his open mouth. "And Betty Crocker?" he yelled after us. "With a Big Red Spoon. How about Mr. Clean....?"

We scattered to our classes, hurriedly trying to forget the dubious wisdom of noon at the nerd table.

Chapter 11

Debbi's Freaky-Fine, Front Flip

Years ago, my new athletic girlfriend, Debbi, along with my dad, Vic, and me, went on a rural Wisconsin hike, seeing and learning all we could in a local, natural paradise. We wandered and wondered through a beautiful, ancient sandstone canyon with a beckoning, clear stream meandering through it, all below steep, rocky walls. On this hot day, we were hiking to get acquainted, and to have some fun together in this wild and pleasantly cool, secluded haven.

Dad and I walked along on dry parts of the lower-canyon streambed of fine sand. Debbi walked up higher, on the steep canyon rim, fifteen to twenty feet above us, as we hiked a bend in the ancient flow pattern. Debbi moved over to the rocky edge of the canyon wall to talk to Vic and me below on the soft, yellow-pink sand. She lost her balance, and fell. In a circus moment, Debbi did a complete forward flip to her solid, two-footed landing in the soft sand next to us. A normal Debbi day.

Nothing was said. Later, Vic told me "That girl has possibilities."

This could have been a life-changing catastrophe resulting in severe injuries and life-long disability or death for Debbi. But with her athletic skills, strength and alert agility, she turned the surprise fall into a successful, impressive front flip...and landed it safely.

Later that day, and through all the years since, when we think of the ramifications of her not executing her fall successfully, we shudder. Brain injury and spinal-cord trauma can affect a person for life if they can live through the initial traumatic onset of their injuries. My profession in rehabilitation medicine brought me in contact with many patients, clients and families of all ages and abilities whose lives were changed forever by a momentary lapse in balance or judgment. For want of not dodging, tucking or ducking in time, whether from a falling tree limb, ski-slope hazard, or stairway in the dark, sharp, capable folks acquired lasting disabilities for which there are no solutions, no cures.

Adaptations and attempts to overcome permanent challenges for the rest of their lives are the only options. Brain trauma and injury, as well as spinal-cord injuries, can and do improve to some extent, but too often their effect is permanent, and can send lasting ripples of adversity throughout entire families and communities.

Things turned out well for us. Debbi and I were married several years later, forty-seven years ago. We have raised two wonderful, adventure-seeking sons to maturity, and to personal and professional success.

And we have lived a lot of active life in between that day of Debbi's front flip and now. Had her instincts and innate athleticism, plus a good bit of luck, not solved her rotational and gravity problems that day, during her two seconds in midair, our lives could have been so different.

We, her nerdy family members, are forever grateful that our well-trained, *and lucky* Debbi adapted and overcame in her moment of unexpected testing. She stuck her astounding front flip and two-footed landing.

Dad and I each scored it a perfect TEN.

Chapter 12

Activity Is Best Anesthesia

We siblings of brothers and sisters who are severely disabled learn to stay busy. We become nerds of necessity, trying hard to keep interested and occupied in our own detailed worlds because the environment around us is tumultuous and fragile.

At least that was my experience. I learned that staying busy helped reduce the pain, disruption and anxiety that I lived through each day. Activity was the best anesthesia. It allowed me to go on.

The isolation and exclusion of siblings who grow up with family members living with severe disabilities was an often unnoticed, unaddressed family challenge in the 1950s and 1960s. As I grew up, any hint of disability, insanity or other differences in a family were enough to cause others in the community to distance from those kids...and their sibs. That was me. Distant, invisible nerd boy.

Keeping busy, always having something new, diverting and challenging to learn and achieve, from pole vaulting to Scout ranks to ski jumping, helped occupy my mind and time. These activities allowed me to push away the loneliness and exclusion I experienced when I engaged in complex things by myself. Amateur radio, Morse code and electronics, along with skiing, skating, hockey, baseball, rock hounding, high jumping, camping, caving and hiking...all of those individual passions absorbed my attention and energies, becoming childhood and lifetime comforts. And I didn't even know it at the time.

They worked. Although I now realize that I was not becoming skilled in *inter*-personal relationships, I was indeed strengthening my own *intra*-personal strengths through learning and personal mastering of difficult things, with focus on many skills and fields of knowledge.

That focus got me through life to this point, and is why I am writing this. Keeping my mind and body active and excited on many fronts, *one at a time*, is my way to not bog down with sadness, depression and loneliness. For over two thirds of my life now, I've not endured this alone: my true hope for others.

The miracle happened along the way. When I was twenty-two years old, Debbi, my lightning girl, my marriage partner for forty-seven years so far, appeared in my life. She changed everything, in all-positive ways. Our enduring love and companionship, our two amazing sons, our grandchildren, and the busy, productive, creative lives we share daily, carry me through life's rumbling roller-coaster ride. When I revert to my solo ways of coping, my wondrous wife and family, with their love, wisdom and support, tolerate me and make each day stellar again.

The anesthesia of worthwhile, valuable activity throughout life has allowed me, and others like me, to remain mentally and physically engaged, productive and well. We can remain healthy through a universe of problems and challenges in our often-long lives, seventy-one years for me, so far. Constant curiosity and attainment of specialized, sometimes arcane knowledge and skills can draw us forward. I will carry on in this same way for as long as this always interesting and fascinating life of learning, adapting and overcoming continues. And I will draw on the deep well and wealth of loving support that Debbi and my family extend to me each day. The miracle continues and grows ever stronger.

Chapter 13

Grandpa Erny And Pal Peetey

"Peetey, Peetey, Peetey!" Grandpa Erny called. The chipmunk's proper name was Peter, intense Grandpa Erny declared, but he believed the little fellow preferred Peetey.

A young, eager chipmunk soon skittered up Grandpa's cabin railing. Erny, eighty-six years old, widowed for two decades, held out an inviting shelled peanut. Peter Chipmunk followed the peanut up to the old man's hand, then to Grandpa's flannel shirt pocket, where a nut was hidden. Peter dived in, placed the peanut into his left cheek pouch, and scampered off to store his treasure in the forest.

Peetey was soon back with Erny's loud call. This time, Erny's grandchildren, seven-year-old cousins, Janice and Tommy, held peanuts. They squealed with delight as Peter ducked into then popped out of their shirt pockets. Peetey packed peanuts into his expanding cheek pouches, then hurried to hide them in his burrow.

Grandpa led the Chipmunk Kids over to his side yard, where he had strung a clothesline at chest height. He held a double in-the-shell peanut at the middle of the white rope and called again. His perky chipmunk buddy stood up outside the burrow, saw Erny's new position, and ran to his friend's big feet.

Strategizing, Peetey looked up to where the new peanut was placed. He raced up the nearer tree and out on the line: a tiny tightrope walker. Peter grabbed the full peanut, pocketed it in one cheek pouch...and *waited*, realizing another may follow. Grandpa Erny rewarded him with one more, then Peetey was back to his underground storehouse.

Janice and Tommy cheered "Do it again, Grandpa!"

Peetey was ready. So was Erny, with his jacket pockets full of peanuts. Together, they showed off their tightrope trick another dozen times. The kids' delight was surpassed only by the amazement of their marveling parents.

Visits with Grandpa Erny were never long enough, and a lengthy drive home also lay ahead. Waving from the back seat, as their 1958 family station wagon crept away, Janice and Tommy watched Grandpa and Peetey do the shirt-pocket stunt again. Grandpa was smiling and laughing, enjoying his little rodent pal who was happy to keep performing. Together, they played for as long as Janice and Tommy could watch. Man and rodent, woodland friends, were co-entertainers.

As Janice and Tommy grew up and had their own families, they spoke often of their long-ago Chipmunk Kids visits with Grandpa, and how he enthralled them with Peetey's chippy antics.

"Grandpa Erny loved Peter Chipmunk," Janice told her children. Tommy agreed, telling his sons of these forest friends and their bond.

In those silly, singular moments, Peter Chipmunk delivered Grandpa Erny from overwhelming isolation and depression. The two characters *were* true wilderness homebodies. Erny's long months and four hundred miles of separation from Janice and Tommy and their families were brightened by the kids' frequent letters and occasional phone calls. Older now, they understood that, way back then, Grandpa's daily social romps with his small forest partner were vital and restorative for lonely Erny. Peetey, companion and comforter, was Grandpa Erny's best friend: Grandpa's closest Chipmunk Kid.

Chapter 14

Hardboiled Mr. Hull

"Rip that flat carcass off the asphalt. Shake it out, then scrape off the big chunks of gravel and bone," he told our group of astonished fourteen-year-old Scouts. We huddled around Mr. Hull, who was teaching us, in great, almost excruciating specifics, how to eat roadkill. He believed you could eat just about anything if you prepared it right. And if you were hungry enough. We could see he *knew* this to be true. He had done it.

Mr. Hull, we were told, was a U.S. Army technical adviser in the early days of the Vietnam War. Now, in 1964, Mr. Hull, age forty or so, compact, dark-haired, solid, was an Army National Guard survival training specialist. He had become a citizen soldier, and was a popular volunteer adult leader with our hometown Scout troop. Officially, Mr. Hull worked for the United States Post Office on our rural mail delivery routes in west central Wisconsin, but we knew he was often gone for months at a time on various military assignments.

Some guys in our troop said he was a Green Beret who knew at least six ways to kill with a stick or a sharpened pencil. I believed it. He exuded capability and confidence more than any Scout leader I met.

But back to the raccoon dinner in progress. First, Mr. Hull picked the rocks and large chunks of other material out of the smashed, flat carcass, turning it over in his skilled hands to inspect and smooth both sides while feeling and checking it with his survival-experienced fingers and eyes. Then he slid the platter-sized slab of desiccated fur, guts and muscle into our boiling campfire kettle, and held it down into steaming, roiling water with a peeled, three-foot chunk of poplar limb.

He turned to us. "Poplar doesn't leach out tastes into the meat. Neither does bamboo." He knew what he was doing.

"Keep it sunk into a rolling boil for twenty minutes," he instructed. "Throw out the water, rinse and check it over. Then boil it again. Do that all at least one more time, and you will have edible food." He continued, emphasizing "You want to do three, thorough twenty-minute boilings."

Quiet, we listened. "If you need to, you could eat a dead snake or any other animal you find in the jungle or forest or on the road. Stick to this method. Don't short-cut. It's survival, not a restaurant."

In an hour or so, we all got to try the boiled raccoon protein that had lain desiccated for many days on the gravel road. Using his Army-issue survival knife from his belt, Mr. Hull peeled and cut out pieces of meat for each of us to try. It was cooked well, gamey tasting but the warm bits of flesh were easy to chew. With some salt and pepper, it could have been pretty good. It tasted OK to us, *outdoors,* on a cold, wet October day during our first survival-training camp out. We were all hungry, ravenous boys, glad we had other supplies in our packs.

"Next time, we'll make turtle soup," Mr. Hull promised.

Amazed and enthralled with what further resourcefulness our instructor-hero might reveal, we retreated to our improvised, rope-rigged-poncho rain covers, spending the night in a hard downpour. We wondered what that next time might bring, awed with the inner toughness and smarts of Mr. Hull. He had survived and taught in the dangerous, war-torn forests and swamps of Southeast Asia in the mid-1960s. We young Scouts had so much more to learn from him in skills, attitudes and self-reliance. And soon, turtle soup?

Chapter 15

Survival Soup From Scratch

Hardboiled Mr. Hull's story continued. We looked forward to learning more about wilderness ways from him. As we woke up the next morning at our Scout survival campout, lightning and heavy thunderstorms caused an early end to our trip. The turtle-soup lesson he had promised happened several months later, by surprise, at our regular indoor Wednesday-night troop meeting in the recreation hall of Our Savior's Lutheran Church.

After our opening ceremony and troop business were handled, in came Mr. Hull with two of his adult helpers. They surprised us, and were carrying a large, square galvanized wash tub containing a heavy, loud load. The scraping, scratching sounds we heard confirmed it was a bulky Alligator Snapping Turtle they had caught in the nearby murky-green lake. Turtle soup was about to happen, right in front of us.

Their efficient process began immediately. One of Mr. Hull's helpers, wearing heavy leather work gloves, grabbed the sixty-pound snapper from its back and sides, then hefted it onto a long, wide wood plank that another assistant placed on the floor. The man in back grabbed the turtle's thrashing tail, driving a large spike through it, into the plank. He held the turtle down on the board. The helper in front poked at the turtle's head with long-handled Phillips screwdriver, provoking the turtle to bite at it. At the opportune moment, when the large reptile extended its neck, opening its mouth wide to bite, the man punched the screwdriver down through the floor of its mouth, pulled its head and neck out as far as possible, and hammered the screwdriver into the board.

With his large military survival knife, Mr. Hull cut through the turtle's neck as close to the head as possible, decapitating the turtle in a flash. He also poked his knife through the extended neck to keep it out of the shell as far as possible. They worked fast in this initial killing, then allowed the turtle carcass to twitch and stop moving over the next half hour, as we talked further about realties of survival food gathering, before they proceeded to butchering.

When the turtle carcass stopped its reflexes of clawing and jerking, the men turned the body over onto its back, the wide, top carapace part of the shell. Then, using a hand saw, they cut along each of the joining parts of tough, strong shell where the under-shell plastron met the carapace. They had successfully severed the junctures of top shell from bottom shell, and now proceeded to cut out the many types of meat.

Mr. Hull and his helpers explained that large turtles like this offered many pounds of edible, tasty meat, just as a large turkey or a goose would. If you knew how to do it, they said, you could get a dozen individual kinds of meat from a turtle, far more than just the white and dark meat we were used to with common poultry. In about an hour, the turtle flesh was cut apart, and separated into different pans for freezing so we could use the many pounds of meat later. All of us in the meeting that night, both men and boys, were captivated with Mr. Hull's skills, efficiency and competence. We were also impressed with his effective, polished methods of leading other men in a complex task.

At a different campout later that summer, Mr. Hull and his helpers brought some frozen turtle meat from that troop-meeting. We boiled it over our cooking fire for more than an hour, making a savory stew with wild herbs and spices. We also browned and roasted some of the turtle meat as you might chicken, duck, or turkey in a camp frying pan, then baked it in our big cast-iron Dutch oven with root vegetables and herbs. Both ways, it was well-cooked, edible, good food. I have never eaten roadkill or turtle since those days, but Mr. Hull taught me how to do both in rugged circumstances. I believe I still could if needed.

It is amazing how you can adapt and overcome when you are hungry, and when you know what to do. Mr. Hull gave us that gift of competence coupled with confidence to apply in solving our other tough, lifetime problems. We could each extend and use those essentials in many ways.

He also taught us how to rappel down a cliff with just a body-wrapped rope, how to sharpen knives and axes to scalpel-quality edges, and how to safely do a defensive backward escape roll to our feet in an attack situation. All of these were things that I became aware of, learned and then perfected over the years because of Mr. Hull. They and he opened our eyes and minds to creative, divergent thinking, and to new, unique solutions, often right under our noses, so that we might further contrive and prevail when situations require.

Mr. Hull was, indeed, a hardboiled, tough character. He was also a smart, peaceable, resourceful and highly skilled leader with his own quiet, canny manner. He became one of my primary models of what manhood means. Throughout my life, I have sought to be like hardboiled, soft-mannered Mr. Hull in all the ways I can. Other guys in our Scout troop from so many years ago did the same. As I write this at age seventy-one, I know his direct, practical approaches to problem solving worked. Mr. Hull empowered us eager young men with confidence and skills to use for the rest of our lives.

Chapter 16

Donut Dog Dunks Canoe

Ice had just left Upper St. Croix Lake a few days earlier in 1995. The St. Croix River, flowing south to Gordon and beyond, was an inviting three-hour canoe trip for our families. Four adults, three boys and one dog, all of us true Northlanders, plus one bag of fresh bakery donuts, headed out for an early April canoe trip on the St. Croix. We had made this fun springtime trip many times before.

So how was it that two humans and their dog were soon swimming for shore? Canoeists getting wet, especially in water that was ice a day or two before, is not a good thing. Here's how it happened.

Our friends, Len and Betsy, with their young son, Jeff, were back at the lake's western "Cozy Cove" shore just in front of Grandpa Favell's cabin site, getting settled in their canoe and stowing gear they would need on our trip. We were all experienced canoeists, and excited to explore the open, clear river in this new spring.

Son Seth and I, along with Shiku, our thirty-pound female American Eskimo dog, had already paddled our canoe ahead to get on the lake and out of the way of the others. We stopped and waited while Adam and Debbi packed their canoe. They launched, then paddled out to join us.

As Adam and Debbi approached us, they asked if we could toss the donut bag to them. We positioned our boats about six feet apart and got ready. I lobbed the tightly closed white bakery bag with recently warm donuts so it would arc and land in their canoe, accounting for the wind.

That's when Shiku jumped for the moving bag. We hadn't planned on that. She leaned hard against our left gunwale at about the center thwart, toward Adam's boat, nearly swamping Seth and me.

Adam reached out quickly for the donuts, as the bag was now far off my intended trajectory. He and Debbi tried to balance and compensate for the sharp, fast, rocking disturbance in their canoe equilibrium. They could not adjust. They went over – instantly. It happened in an eye blink. I can still view the slow-motion mental movie.

As you can imagine, they gasped, splashing and thrashing in the cold water. We saw the surprised fear in Shiku's eyes as she surfaced, a sleek, soaked, wet-rat dog; so skinny, so scared. Shiku dog-paddled to the shore a few yards away, shook off, and watched her humans perform.

Debbi and Adam kept their composure and swam for shallow water. My shouting "She's gonna die!" toward my water-wary (and water-soaked) wife may not have been all that appropriate. She was fine.

Adam and Debbi simply swam a few strokes toward shore, got into waist-deep water, and found firm lake bottom to stand on. They easily walked into the grassy lake bank. Shiku, now content and mostly dried off, sniffed around on the shore, likely still thinking about those donuts.

To everyone's good fortune, the day was sunny, with a strong, warm south breeze. We retrieved the flipped canoe and pulled it in. Our boys and I lifted and drained it, setting it upright, gently beached on the gravel bottom near shore. Debbi and Adam had wrung out their clothes and socks, and recovered all but one shoe. We were happy, laughing, and ready to go again in minutes.

Through all of this, our friends were still back at Cozy Cove getting set to launch. Things take longer with a four-year-old. Their attention was diverted from the feral festivities we were providing on the lake, out of hearing range for them. We had just drained the canoe and were drying off on the turf bank when they paddled up close enough to see the aftermath.

Len, Betsy, and Jeff registered little surprise as they got closer to us, saying it appeared to them as normal for the Kings, simply activities they expected of us: a quick cold-water dip and canoe swamping, just parts of our annual spring rituals, they assumed.

Humans and canine were soon warmed and willing, so we headed downriver, expecting our usual lazy trip. This day, however, the headwinds from the southwest were fierce. Our "easy paddle" took hours of hard stroking to finally dock at the Gordon ranger station.

We gained insights that long day into why Native and European voyageurs, traders and explorers poled their way *up* this part of the river, often in low, flat-bottomed boats known as *bateaus*. Going upstream against flowing water currents would have been *way* less work, we figured, than bucking the warm but hard, pernicious, southerly spring winds throughout our entire trip.

Overall, family lessons of resilience were learned that day, with great memories also of Len, Betsy, and Jeff, our tolerant, good friends. By the way, the donuts and bag sank.

We never found them.

Chapter 17

TRUE 1890s FARM STORY

~~A Smart Dog~~
Handwritten in pencil by Great Grandmother Angela Haste Favell, circa 1893

Some years ago we were living on Upper Lake St. Croix at Solon Springs, Wisconsin, where we had a partly wooded ten-acre tract. On this we built a barn and converted most of the space into pasture for two cows. A friend gave us a puppy, half collie and half shepherd that grew into a very intelligent and useful dog.

Going with his master to the milking was a chore our Jack greatly enjoyed and never would miss. He would drive up the cows, wait until milking was over, then return home carrying whatever article his master might entrust to him. Only one thing he enjoyed more than this milking trip. That was to hunt rabbits with our son who came from the city week-ends.

One Saturday after an early supper our son got out his gun and said to the dog, "Jack, bring my hunting boots and we will get a rabbit." Jack acted a little excited but confused for a moment. Then, instead of going right to the closet for the boots as he had done so many times before, he bolted out the kitchen door and ran up the hill toward the pasture. We were still puzzling over his unusual conduct when suddenly he reappeared, ran to the closet, brought out the boots and deposited them at our son's feet, thus announcing his readiness for the hunt.

Shortly after the hunters had gone, my husband remarked that it was nearly milking time and that he would start to drive up the cows as his helper had gone hunting. Imagine his surprise, on arriving at the barn, to find the cows already in their stalls. Jack had been there and done his work before going hunting.

(Signed) Mrs. Angela E. Favell, Superior, Wisconsin

Note: This true story is presented here verbatim from the original, and transcribed without editing. It was hand-written about 1893 by Angela Haste Favell, my great grandmother. She typed it circa 1920, and lived more than 102 years, passing on in 1953, in Superior, WI. In the later 1800s, about two decades after Thomas Favell, my great grandfather and namesake, returned from four long, dangerous years in the American Civil War, he and Angela homesteaded in a small, quickly built log shelter in forestland east of Upper St. Croix Lake. Later, they cut and hauled logs from that property across the ice to their new land mentioned above. They used those logs to build *Shore Acres* and *The Maples*, their farmhouse and guest cottage, around 1890-1900, homesteading and farming this time on the west side of the lake. Thomas and Angela spoke of *Shore Acres* as the first house built on the west bank of Upper St. Croix Lake in Solon Springs, WI. We live on the part of their homestead where this true story happened about 130 years ago. - twk

Chapter 18

Power Of On By!

We can learn a lot from sled dogs. Our totally *un*official, *un*certified, unregistered, and unruly homestead dogs, Scotty and Sonja, are eager, trained Sheltie/Border Collie-mix power pullers in their own way. They may not be professional sled dogs, but they *do* focus on the essentials. They know: *Gee*, go right; *Haw*, turn left; *Hike*, charge ahead fast; *Whoa*, stop; and *On By*, ignore irrelevant distractions.

Most of us who are pulling hard in our lives and jobs, regardless of our own pedigrees or lack thereof, are pretty good, in our own ways, at the first four of these. Our language has many social metaphors for these concepts, going well beyond the physical actions described.

But that last one, *On By*, is still something we all can learn. Getting our work done without being diverted is often difficult. Tuning out distractions and disruptions in our lives, day and night, is hard to do, and seems to get harder the older we become. Acting upon *On By*, meaning not worrying, not fixating needlessly on concerns about family, work, politics, and so on, gets harder to do as we add to the worry-weight on our individual dog sleds of life.

Pulling our heavy loads takes its toll on us. Curiously, the growing popularity of expensive, designer stimulant beverages and other pick-me-ups we use seems paralleled by growing attention to the array of relentlessly advertised sleep aids. Our custom as a society appears, increasingly, to be: animate ourselves with some products during the day, then try to calm ourselves (and "quiet our active minds," as the ads say) with other heavily advertised products at night. That good night's sleep still seems ever evasive. The use of uppers during the day and downers at night, unfortunately, has been around a long time. Now our ups and downs seem more and more mediated by mass-marketed products and brand names, be they caffeine sources or prescription sleep medications.

We are going both ways at once.

Maybe the sled dogs and their drivers have it right. We all need a firm reminder sometimes to let certain things go *On By*. When we have done what we can about our concerns at home or work, or with our health, family, or finances, we should keep our focus and keep moving ahead – just like the sled dogs do. We can continue *On By*, and not let those worries, distractions, and disappointments lure us off the trail.

Yes, sled dogs can teach us a lot about doing the best we can in the moment, with what we have, and letting the rest of life work out as we run our course. *On By* may be particularly useful to remember at this point in our lives. Not "Hike it up!" press on, pull harder. Not "Whoa!" stop trying or caring. But *On By*, give ourselves permission to keep on going, while we let distractions and memories of slights, errors and imperfections remain in the background of our minds and lives.

Sometimes it's good just to pet and comb your dogs, while you sit on the deck near the resilient, enduring old-growth forest, and let the world work itself out around you. Sled dogs are mighty smart.

Chapter 19

What Party?

"So, what party do you belong to?" People do ask me that occasionally. I am never quite sure what to say. As a non-card-carrying member of any political group, it seems I am mainly a loyal Lake Superior region booster, part of my imaginary Norway and White Pine fan club here at our forested home and studios.

But a member of a political party? No.

Over the past several years, I have had time and energy to ponder the many-faceted, complex, always-vexing political issues we face. These are perplexing, but it is easier to think more clearly and deeply as I work outside in this ancient pine forest. Now, I must often consider the world from many perspectives, attending to what truly matters the most to me.

Here are some of my thoughts these days as a rural-dwelling, lake-loving, forest-friendly resident of far northwestern Wisconsin who thinks constantly about this planet, continent, nation, and state...and where we may be heading with which kind of leadership.

In many ways, the Green Party speaks to my heart about our

environment. They appear to share in and promote my reverence for our natural world. We live in a little clearing in an old-growth forest near Upper St. Croix Lake, in rural northwestern Wisconsin. Understanding and wisely stewarding our natural world, while blending with it, are important and essential.

I also agree with the Democratic Party about working for the common good of all, and helping residents and citizens to live better lives through life-long learning and public-policy initiatives. To me, Democrats seem to care about our environment, people, health care, education, and several other issues in ways I believe to be important.

But I find a lot of value in Independent ideals of making my own decisions, and being able to choose among candidates and issues on my own. Deciding and voting as we each believe best, not because of a party platform pushed on us by political organizations, is central to our way of thinking in rural North America. Independent, objective thought, action and responsibility, are central to individual and societal success.

Libertarian ideals, as I understand them, of "educate, don't regulate" and of keeping government out of our personal lives, have considerable merit, too. A government that governs informed, literate citizens least is indeed best, to paraphrase Jefferson. I'll bet we all have much more libertarian in us than we may admit.

Republican belief in individual efforts and energies being rewarded with individual gains appeals to me, as well. I work hard, always have, and so have all my family. I believe nothing confers dignity on a person in this culture more than working to take care of yourself and your family, to achieve your own dreams and to make your own way in this complex world, as you create livelihoods for others.

Unfortunately, there are serious downsides to all formal political party ways of thinking. These trouble me.

"Green" in this country is too often represented with behaviors and ideas synonymous with another five-letter word: *f-l-a-k-e*. I wish that were not the case.

Democrats seem to place much emphasis on everyone sharing, on giving too much away, with not enough value on the innovative, entrepreneurial spirit of motivated individuals which can change lives.

Independents can be criticized for not aligning on issues, for creating policy checkerboards in governance at all levels. Non-alignment can result in significant non-effect and chaos. Libertarians propose a society with ideally-minimal regulation, where we all act wisely on what we each "know" to be best, and everything works out well. This seems unfeasible on a large scale for any human world in which I have experience. Think recent life and events in Somalia.

Republicans too often seem to fit the stereotype of "born on third base, and think they hit a triple." The "I got mine, too bad about yours," winner-take-all mindset is destructive and repugnant in my thinking.

So, as you can see, I am unsettled and conflicted. I believe all of these political conceptual frameworks have merit. But they all also have glaring, disturbing deficiencies. At this three-quarter-way point in my life, I am simultaneously in agreement with, and revolted by, portions of each of the above parties' ideas and actions.

So, next time I'm asked what party I belong to, maybe I'll just say "I'm a planet-loving, intelligently-alive *Demo-Indepen-Liber-Republ-crat-dent-tarian-ican*." That'll settle it all for good, I'm sure.

By the way, next time an election comes around in your area, even a small, seemingly unimportant one, study up on it, think about it thoroughly, discuss it openly, then do that truly patriotic, so-important exercise of your rights and duties as an informed citizen: *VOTE!*

Chapter 20

Isle Royale Marathon

Debbi and I celebrated our first wedding anniversary in the summer of 1976. As a newly married couple, a real team, she and I were eager to hike and camp, geeking out with abandon, on our beloved wilderness backpacking trails of Isle Royale National Park, in northwestern Lake Superior. This particular island adventure took us on many turns, literally and figuratively, over rugged, wild island trails, and had us completing our own *unintended* Isle Royale Marathon Hike.

On our first day out, we walked from the northeast Rock Harbor area, where we unloaded from the Wenonah boat we had taken from Grand Portage, Minnesota. We hiked the easy one-day trek to McCargoe Cove camping area, about fifteen miles, near the northeast coast of the island. Our plan was to relax there for two days or so, swimming and fishing, then make our way gradually up to and along the challenging Minong Ridge Trail, which parallels the other high-country, scenic backbone of the Isle: The Greenstone Ridge Trail. We expected to slowly hike our way incrementally, as weather and interests directed us, back to the Washington Harbor/Windigo campground on the far western side of Isle

Royale, a thirty-some mile hike, spread over several days, to meet our return boat pick-up at Washington Harbor.

Thinking we would stop at Chicken Bone Lake campground, or make a wilderness camp part way up the rugged Minong Ridge, we started out into the interior island trails the next day. Mosquitoes and an abundance of hikers on the island that week kept us moving. Soon, we were up on the Minong Ridge Trail; well on our way up steep basalt ridges to the top, and among fewer people and biting flies. Moose, wolves, foxes and beavers were the primary large mammals on the island. No bears, no skunks. But prey-predator balances change. We kept moving forward with no interruptions other than occasional rest and water breaks, hoping to see and hear some of our wild neighbors along the Minong Ridge peaks and valleys, as we walked on with full packs and water bottles.

To the southwest, on the parallel Green Stone Ridge Trail, was the Ishpeming Fire Tower. We would see it from the Minong Ridge on our hike toward Windigo, so the mid-island tower region seemed to be a reasonable goal for the day. A wilderness, off-trail camping place could be found about halfway along the east-west spines of the island, where we could sleep, cook, boil water, rehydrate and rest. The gradual hike toward Windigo a few days later could unfold just as we had planned, with stops along the way as weather and our trail interests directed us.

That plan would often be viable, except that rain is frequent and heavy on Isle Royale. Hard rains began. The sky opened, and we got soaked. Our plans changed. We stopped and put on our full rain gear, then hiked fast up to a resting place in the low, leafy forest, out of sharp lightning and rain for a while. We were then approximately north of the Ishpeming Fire Tower, a few miles away, so we made a quick one-pot meal of scrambled eggs on our small, dry-wood cooking fire, boiled lake water we carried, and drank hot raspberry mix as we plotted what to do next. That lunch has proven to be one of the most memorable, important meals of lives. Our nerdy need for nutrition, even in the middle of wet wilderness in a fierce storm, guided our actions, and was, indeed, the right choice at that time. We were rested, ready and reenergized.

The surprise electrical storm and brisk uphill hike had caught us partially prepared. We were simultaneously drenched and dehydrated from outside rain and our inside sweating. Our thirst and fatigue were abated with the wonderful small, warm lunch, rest and liquids. In the flashing lightning and thunder, we contemplated deeply: Do we stay and set up camp wet, hoping we would dry out by the next day? Or with our drinking-water supply now depleted further, should we keep going, trying to get closer to Windigo, and hoping weather improves?

We decided to go for Windigo. In full rain gear and uncomfortably wet already, we were now fueled and restored from lunch and liquids. We hiked the entire remaining distance of about twenty-seven miles into the Washington Harbor/Windigo Campground, finishing to our goal *in one determined push* that long afternoon, and arriving in early evening. Our total map distance hiked that day, under full packs, was over thirty-five miles from McCargoe Cove to Windigo. Trail-map markings do not reflect the hundreds of meters of ups and downs, ridges climbed, and beaver ponds forded, waded or detoured by bushwhacking around downed trees. So we covered more than the marked distances, and were relieved to finally be in camp. We could relax, dry out and warm up.

Our final push to camp turned into a marathon effort for us. Miles of rugged, narrow, often precarious trails, with changing up/down elevations from lake level to ridges and back many times, full packs, hard rain, lightning, and fast hiking in back-country muck and water over our boots...all were a tough challenge. We were stupid and bold, but now pleased, realizing we were fortunate survivors of a day-hike that went out of control in tough territory during a powerful storm.

We had made it into camp safely: tired, blistered and thirsty, true, but rejoicing to have made it in one determined haul. We would be able to meet our return boat on time, with some fun days of relaxation and exploration in between. Although neither of us had, or has ever since, entered a posted, commercial marathon, we consider our informal Isle Royale Marathon Hike on that flashing, thundering, pouring day to be plenty good enough for us, for life.

Over our forty-seven years since that trip, our frequent and numerous five- and ten-kilometer and ten- or fifteen-mile running and skiing races have garnered us T-shirts, cups, medals, and many friends. Each day we run, hike, or ski, plus tend our farm, forest and critters, first covering our neighborhood three-mile loop, a thousand miles each year; often 15,000 steps or more per day, if you wish to calibrate activity that way.

Isle Royale strengthened our spirit, our resolve, and our bodies. It proved to us early in our married lives that we can accomplish, probably even exceed, our plans when we do it together. The physical marathon hike we inadvertently completed on the Isle in our first year together has proven to be metaphor, encouragement and catalyst for challenges and triumphs we have encountered as a team throughout our four-plus decades since, with many more decades to go, we trust.

Our first light, early-morning outdoor trek each day of each year is our continuing, bold celebration of adventure, discovery and each other. We are grateful for active lives each day, and for those trail-side scrambled eggs that helped power us to our distant destination through heavy storms. This true tale is a metaphor that keeps us ever thankful.

Chapter 21

Full Contact Smelt Run

Bears are passionate about fish. So are humans. And both species enjoy smelt because the small fish are easy to get. Many late-spring nights, large schools of Lake Superior smelt are bountiful, easy to scoop out with nets or claws, along the shore or in streams and rivers entering the Big Lake. Smelt are among the best yearly snacks that nature offers.

In April of 1973, I was living in Thunder Bay, Ontario, and eager for smelt season to arrive. The mouth of the Current River, just north of the Port Arthur side of town, was an excellent smelt venue, I had heard, so I got my gear ready to go. My dip net, two-man seine, pails, cooler and a borrowed galvanized laundry tub would work well.

As an experienced, well-equipped *caver*, I also had my rugged flashlights, and my helmet-mounted carbide miner's lamp, which produced a bright, piercing beam with its live, two-inch acetylene flame. The brilliant lamp could enable me to see far down into the dark lake and river waters at night to find surging schools of smelt.

Along with several friends from the area, who also brought

their gear, we drove to the Current River juncture with Lake Superior, a short trip, just before midnight. We checked out the beach and sea walls along the river entry for smelt activity. It was hopping or rather wriggling with smelt swimming through in large schools. Fisher folks were hauling smelt up with full dip nets; almost shoveling the little fish into their tubs, pails and coolers. We filled our containers quickly and were packing up after just an hour to go home for early-morning smelt cleaning, as we'd prepare our thousand-plus tasty smelt for freezing.

While I began to haul gear up the lake bank to our van, a small, frail woman with a high, gray beehive hairdo of the 1960s, came up to me holding out a plastic shopping bag. "For my cat, please? Please...?" she asked in French. My dip net was in my left hand, so I set down my other items, walked a few steps to the bank of the Current River, still teeming with schools of wrigglers, and scooped out a load of the small flopping, silvery fish for her. I leaned over to carefully dump them into her plastic bag, filling it fast. The bag gyrated on its own.

Next, this petite, older woman stepped in close, and looked up at me. "Merci! Merci... Pour mon Chat!" (*Thank you. For my cat.*) She spoke with great sincerity...just as the hot flame from my helmet lamp ignited the top of her beehive-stacked hairdo. Her hair was blazing.

Hair-sprayed dry hair burns rapidly. With my wet fishing gloves, I patted, pounded her hair and head to put out the flames. A saucer-sized hole had burned into the top of her hairdo, but the fire was out. Her skin and scalp had not burned. Just an upper circle of her tall hair.

This all happened before we could grasp the disaster we had averted. She walked off into the night, startled, I am sure, pleased to have her smelt in her bag for her cat, or her meal. I am also certain she was dismayed at why this helpful, strange young man was whomping her head so vigorously. In the midnight darkness, the crowd of eager smelters, and the confusing swarm of languages, she slipped away. I tried to catch up with her to assure she was uninjured, but never found her, and still wonder about her reaction when she saw herself in the mirror. Lake Superior smelting can be a full-contact sport.

Chapter 22

Kruska Lunch In Reykjavik

Exploring sprawling Reykjavik, capital city of Iceland, on eager, jet-lagged feet, was our unforgettable adventure during our second day there. As we hiked across hills and residential neighborhoods, then through commercial areas and national sports venues, wooded lanes, and into old and newer portions of the urban center's arts district, we saw and heard so many new things: tributes to elves, fascinating outdoor sculptures, the Greenland Sea in view with ancient volcanoes in the distance, and helpful, friendly people at every stop who all spoke English better than we did. We felt at home in this new city.

With morning light coming on stronger by about 10 AM on this late October day in 2019, we hiked till we dropped, or at least until we knew it was noontime. We were getting hungry.

As we strolled past the inviting front gates of the Reykjavik Zoo, we stopped in to ask the desk clerk at ticket sales where she would recommend for lunch. She suggested we walk back up the hill, take a right at the top and look for eateries. She further explained there were several choices in a small strip mall that we would see as we ascended the scenic hills. The expansive harbor, plus Mount Esja (Esjan) and distant snow-topped mountains were majestic.

As we topped the hill, our foodie instincts drew us to the small shop emanating great bread aromas. KRUSKA the sign read. Yes, Kruska was the perfect, correct lunch restaurant for us to stop in that day, on our trek through hills and back streets of real Reykjavik. The sight and smells of fresh bread, unique soups, and complex salads drew us in.

Kruska's fresh, mildly hot carrot soup of the day, along with their unlimited bread with deep-yellow, creamy Icelandic butter, and fresh hummus caught our attention. Our helpful server suggested one of their varieties of salads, and did we ever choose the right one. It came at the end of the meal, as it should, and was a festival of red cabbage, orange carrots, light green and sweet jicama, dark lettuce, beets, cucumbers, a bit of onions and more, highlighted with a unique dressing new to me: a festival of colors, tastes and textures. Truly, the best salad I have ever had. And our smooth, aromatic carrot soup, along with their rich butter, fresh homemade breads of several types, combined with their tasty hummus, were also excellent. The comfortable, chatty ambiance of Kruska staff and patrons enhanced our already-great experience in lunchtime joy.

Back to that salad. Debbi and I focused on it, examining it up close, as we noon-hour nerds picked it apart, ingredient by ingredient, reverse engineering the complexity and tastes of this ingenious lunchtime creation. We took pictures of it. I am examining them now as I type to be certain I acknowledge each of the amazing ingredients in this memorable, new-to-us combination. I have not covered them all because I have never had a salad with such variety, texture and taste.

Maybe it was the hike, and our relaxed, laid-back vacationers' vibe at this small café in beautiful Reykjavik on our first trip there, but that was a meal, and especially a salad, to savor in memory for life: a lunch experience for folks who love salads. Any one of the ornate, friendly Kruska cafes in Reykjavik will be on our noon priority list for lunches as soon as we can get back to Iceland.

Chapter 23

Samoyed Raids Valentine Cookies

For a few years, each visit I made to the local humane society shelter resulted in my bringing home another dog. My wife finally cut me off.

But one of our rescues, "Snowball," our eighty-pound, all-white Samoyed, moved in with us and soon grew to love her pampered life in our small home. She took over from the first day with us, having been dumped at the county animal shelter, referred to as "dog pound" at that time and place. Her former owner resented her sleeping on his septic-tank drain field during winter for warmth. He also believed she was too friendly, too wanting of the companionship with her people. When I found her, I signed her out, let her jump into our car, glad to have her, and took her home, where Snowball soon was in charge.

We had a lot to learn about adopting a large, adult dog into our new household. Snowball was gentle, beautiful and easy to be with. She readily lived right inside our home with us, but ate and drank, and pooped a lot around our small yard. No problem for us. We accommodated her quirks and needs, delighted to have her with us.

Early in our relationship, we found that Snowball loved to pull us on skis. She knew what to do. We got a proper skijoring harness for her, soon enjoying fast travel with our cross-country skis on snowy trails and frozen lakes, with Snowball as our all-white locomotive.

Debbi and I were both working in those BC (Before Children) Days, so Snowball often got to stay home alone. She was house trained and a good citizen around our home. Valentine's Day was approaching, and my critter-loving wife, also the consummate master baker and culinary artist, made several batches of fresh, heart-shaped, from-scratch sugar cookies. She decorated them with pink icing and had set them out to cool on top of our elevated, kitchen-island countertop as we left the house for a short errand that day.

When we returned, coming in the front door, there was our dear shaggy fluff-ball puller, Snowball, up on the bar stools near the kitchen island. She was spread out across the stools and countertop, gorging on cookies, even mopping up crumbs among the array of tasty valentine treats. With great determination, she kept on devouring them as we came in the house. All we could do was laugh as we tried to discipline her. It was such an outrageous sight and humorous scene that we just took her down and tried to salvage what we could of the cookies. Some were untouched, but Debbi baked more. We filed the experience under getting to know Snowball, our new housemate. We were glad she had intense passion for treats that we also liked.

Years later, as we had our sons, Snowball went to my parents, who loved her for the rest of her long life. She pulled my dad, Vic, on skis through days of snowy adventures. Maddy, my mother, collected Snowball's long hair, and had it spun it into yarn for a beautiful, knitted hat. These thing happen when a lovely dog and willing humans meet and fall in love. Thanks for all those great years, Snowball. You are still loved as our wonderful, fun partner.

We think of you often, especially on Valentine's Day each year. Heart-shaped cookies with pink icing are always our wonderful remembrance of our dear friend Snowball.

Chapter 24

Treachery And Steel Head Trout

True outdoor adventure stories are always happening for us fortunate residents and visitors in the Lake Superior Northland. Friends and I decided to fish the large Steel Head trout that were running from Lake Superior into the Neebing River, just a few blocks from where I lived in Thunder Bay, Ontario.

To catch these powerful trout in that river, you slide a red salmon egg onto a small hook to just cover the barb. The hook is tied directly to your monofilament line. No leader line is used because it alerts these wily fish, in spawning mode, to your ploy. You cast your baited hook out into the swift-running, narrow but deep and swift Neebing, then let the bait settle to the bottom, keeping your line slack as you walk downstream. The bait bumps along on the riverbed if you do it right. When you feel the slightest tug that is not a bump, you set your hook with a quick snap of your line and get ready for a fight to reel in your trout.

About twenty men and older boys were fishing on each side of the river that day at the place we positioned ourselves. Lots of movements and excited talk went on among the group as several big fish were landed. Some catches were amazingly large, and we noticed that a few fish were gutted and field-cleaned on the spot. These fish parts, along with discarded baits and human garbage left each day, messed up the area, likely creating attractions for bears and other critters who were becoming active again after a long winter.

Admirably, most folks cleaned up well. Some did not. On that early morning, as we fished just after daybreak, we were among the excited, intent group on the Neebing doing our best to honor the river and shore. And catch trout. That is hard to do. The atmosphere was tense.

Loud, angry shouts suddenly erupted from both sides of the banks down river. As we watched in disbelief, a scuba diver was gearing up to submerge a block or so downstream from us. He put on his dry suit, tank, mask, fins, gloves and other gear, then waded into the river and ducked fully under. We could see his flipper splashes and air bubbles as he swam upstream, underwater, coming toward our agitated crowd.

In the largest impromptu response of group aggression I have ever witnessed, men and boys fishing on both sides of the narrow river began to hurl softball-to-soccer-ball-sized rocks and concrete chunks from the anti-erosion riprap piled along the riverbank. These guys were mad, aiming directly for the diver whose bubbles kept moving upstream. Harsh language and threats abounded, as the situation grew dangerous and belligerent.

Our group were all cold, wet and hungry by mid-morning, and disgusted at the behavior of these "sportsmen" and diver. All the rowdy agitation left us with no catch, so we left, and never found out what happened or why. Events that morning, decades ago, became another adventure from the shores of western Lake Superior. What had we witnessed? Research, curiosity, malice...stupidity?

Chapter 25

Simple Barrel Stave Skis

Two simple barrel staves started it. When my dad was a young farm boy, two old, slightly modified barrel staves began our lasting family fascination with skiing. With the intense focus and determination of an eager, energetic, lone child, growing up with older parents, Victor learned to balance and "ski" with his bent-wood, toe-strapped curved staves on the gentle front slope in front of his family farmhouse. He became proficient at it, such that his father, Gideon, made wooden skis for their immigrant family. They struggled financially, so making up clever, free fun from whatever they could craft around their small farm became essential.

Victor's father, Gideon, came to Hayward and rural Rice Lake, Wisconsin, from St. Alphonse, Quebec, Canada, working as a skilled logger in the Wisconsin timber industry of the late 1800s. He understood the importance of balance, skilled confidence, and homemade fun.

Gideon was handy at improvising in the forests, so he fashioned wooden pins to secure homemade leather toe straps across two curved barrel staves for his little son, *"Veek TOR"*. Dad said his father, my grandfather, always pronounced Victor's name that way in Gideon's French-Canadian accent.

Victor honed his skiing skills on that small front hill at their rugged, hardscrabble farm, and got pretty good at it. He caught the ski bug, which he passed on to me, and I recall that he never owned real skis until he was out of high school and working. When Victor finished his pharmacy degree and license, he and my mother, Madeline, were able to afford better ski equipment in the late 1930s and early 1940s. After World War Two, their emerging prosperity allowed them to outfit their growing family and pursue their downhill passion. They taught me to ski when I was about a year and half old in 1952, using small commercial wooden skis (Lund brand) with simple leather toe straps.

Soon, ski jumping on easy farm-hill snow bumps, and skijoring behind my pulling parents, as well as behind cars or trucks on lake ice, became part of normal, growing-up life. Eventually, downhill alpine skiing and racing, backcountry trekking and camping, and cross-country ski touring became my obsessions. All of which derived from the reality that, other than walking and talking, the first thing I learned to do with intentional skill as a toddler was ski down neighborhood hills with basic equipment: two long wooden skis with leather toe straps, snow boots, and no poles. Nothing fancy. I just slipped toes of my winter boots in the straps and pointed my skis downhill, seldom falling.

Balancing on all types of terrain while skiing fast over snow of all varieties, working with gravity, became the most interesting thing I have ever done. It still is. As of this writing, I have skied for 70 seasons, and am looking forward to 70 more. Hey, a guy can dream.

From my parent-guided backyard start, all variations of skiing have been my delights ever since. I inherited and built upon their intensity. From gliding straight down steep back farm hill headwalls in deep snow (or two worn tracks, one for each ski) then hopping from jumps and bumps on the way down, to downhill and slalom racing in high school and college, National Ski Patrol ranks and directorship, ski instructing, and NASTAR gold medals in giant slalom as an adult, I have followed my ski bug in every way possible that is safe...and some that are not.

Cross country, backcountry and alpine skiing, ski jumping, and many days of fast, hair-raising skijoring behind trucks, cars, snowmobiles, dogs, you name it, in all temperatures and types of snow, yes, I have enjoyed them all. And I have been to some awesome, challenging places and met incredible, durable people in my lifelong ski quests around North America.

Skiing jumped forward to our next King generations, too. My wife, Debbi, learned quickly and well as we were dating. Over our 47-plus years together so far, she and I have passed the ski bug along to our boys and their families. We still ski in a variety of ways on every day possible in our busy lives, updating our precious gear as we can afford.

With our primitive shepherding adventures of the past decade, our versatile backcountry skis and poles, along with our balance skills, strength, and conditioned endurance from hauling heavy loads in deep snow have become the reliable methods of moving essential water, feed and farm gear on our sleds when roads are blocked. The fun and functionality of skiing continue to be challenging sources of delight and satisfaction for us. We ski out of necessity many days each year when no other means of transport works at our remote, rugged farm.

Yet, the thrill of fast, downhill high-mountain skiing, through cold wind and powder, or over hard pack snow, will always be my supreme joys. Back country and cross-country skiing are certainly useful and fine, but speeding my razor-sharp giant slalom skis through gritty down-mountain turns, resisting G-force compressions of each directional change...well, that tops them all. And keeps me addicted.

When I was sixteen, on my high school alpine racing team, I wrote an enthusiastic-nerd poem for a ski magazine. My verse ended with "Lord, if there's a Heaven, let it be on Skis!" Words still so true.

Chapter 26

Lunch With Colonel Sanders

Yup, I once had lunch sitting next to Colonel Harlan Sanders, of Kentucky Fried Chicken (KFC) fame. He, along with about a thousand Boy Scout adult leaders, Scouts and other volunteers, gathered at a large outdoor picnic area in an open, grassy field of a metropolitan recreation area and park big enough to accommodate our huge crowd.

The Colonel was the featured speaker for this national Scouting event held in suburban Chicago in the spring of 1975. Dressed in his full, all-white colonel suit, he looked just as you would see him on TV. I was seated with adult leaders at a different table, many rows out on the lawn, some distance from his head table of dignitaries. As I recall, he was uncomfortable in the intense noon sun where he was originally placed, so they moved him to our table, to the folding chair right next to mine.

We all got acquainted and developed a warm appreciation for our guest. Colonel Sanders (Harlan to us at our table) was a remarkably nice guy. He was polite, affable, talkative, engaging and cool. During our lunchtime conversation, at some point in our chat before he spoke to the whole crowd, he turned to me and said "Tom, I never made a nickel until I was sixty-two."

His theme of never giving up was central in his speech that day to Scouts and leaders, with emphasis on keeping on trying until you succeed. Harlan tried and succeeded in his own life, as we all know. He impressed us with his personal message, his intense drive to succeed, and helped us envision our own values and goals.

The Colonel spoke of his many business struggles that day, and especially connected with us at our folding table on the grass, creating lasting good feelings in us about him and KFC. He was a role model for us to follow. Our picnic lunch that noon probably was Kentucky Fried Chicken with all the fixings, courtesy of Colonel Harlan Sanders, but I don't recall. I do remember the intelligent, affable, highly capable man I sat next to for that short while, and recoil at the demeaning depictions of Colonel Sanders which we often see now in commercials. Harlan Sanders was much more than an advertising comic or cartoon caricature.

The Colonel was a focused man who pursued an idea and dream that he refined, developed and stewarded into astonishing global success. I am honored to have met and become briefly acquainted with Colonel Harlan Sanders on that sunny spring day among fellow Scouts and leaders in a suburban Chicago park, a memorable lunch break.

Chapter 27

Park Workers' Disney Spells

In 1966, I was just old enough to be a useful, legal driver, so my parents took my sister and me on a road trip throughout the U.S. Southwest, with several stops in California to visit relatives and family friends. The four of us spent a day at the original Disneyland in Orange County.

Because I was sixteen and believed myself too cool to be seen with my parents and older sister, I explored the huge theme park on my own. My day filled with giant squid, rockets, submarines, talks with kids I met from all over, and encounters with skipping, dancing, whimsical, large-headed animal characters, including a bear or two, all come to life on the streets, from the cartoons I had seen on TV. The Wonderful World of Disney program, in full color, was on NBC each Sunday night, and my family watched it nearly every week.

My lunch break at the park's Frontier Land Cafe was a uniquely memorable part of the visit. As I ordered and ate my tuna salad sandwich at a table on one side of the cafe, I watched a loud, growing group of lunchtime, headless familiar characters in their still-whimsical costumes eating lunch in a corner alcove of the room.

These famous licensed characters, I began to realize in my still-too-innocent mind, came to life for us tourists through the hard work and sweaty animation efforts of real women and men actors in hot, furry, feathery, leathery outfits, with removable heads. These were breathing, hungry, thirsty people in mouse, bear, chipmunk, and fairytale suits.

An entire group of them spread around a couple of tables not far from where I (the only other customer in the cafe) was sitting. They were having a lively discussion about their wages, at around $1.65 per hour, as I recall. Their personal animation increased as they conversed about their pay and work. Some were not pleased. During that summer, I was making $1.25 per hour on the city street crew back home in small-town Wisconsin, painting yellow traffic lines and parking lanes on 100-degree black top, so their hourly wage sounded pretty good to me. And their work looked like way more fun.

As a first timer at the "Happiest Place on Earth" I was stunned, dismayed at the lesson I was getting. Realities of pay and benefits actually here, too. These otherwise so-happy, bouncing, always-affable characters took off their headgear and grumbled in the corner, just like the crew I worked with on the hot pavement. Yes. It was happening.

I came away changed, realizing that even the most famous, glamorous folks are striving. We all eat, work and try to get paid a fair wage, pursuing our sometimes-cartoonish lives as best we can, never able to duck the ever-present realities of earning a living.

Even at that amazing, happy venue in 1966, the live characters broke for lunch, and with passion and intensity, discussed their unique employment. Then they used the washroom, put on their costume heads and went back to work. Even the happiest place on the planet had real people working, with real concerns amid all that surrounding joy. Their Disney spells taught me a lot during that lunch time.

Chapter 28

Hiking Lake Ice 119 Days

My wife and I are determined, durable skiers and hikers of lake ice, finding the silent splendor of deep-winter adventures on Upper St. Croix Lake to be our favorite days of the year: quiet, peaceful, pristine.

New glistening snow sparkles in early morning starlight, and in bright sunlight when we trek the ice later in the day. Tracks of wolves, coyotes, deer, and otters hint at many new stories to discover. Eagles and ravens in flight guide us on our unbroken white way.

In the calendar year of 2013, Debbi and I completed 119 lake-ice hikes, each on different days in Solon Springs, WI, traveling on solidly frozen Upper St. Croix Lake. Family members, along with Sonja Snow Dog and Scotty Boots, our sister-brother Border-collie-mix sheepdogs/sled pullers, frequently accompanied us. We walked or skied, often on local North Country Trail links, out to thick, safe lake ice, then picked our hiking or skiing routes based on wind intensities and directions, and snow temperature and firmness on that given day.

Mid-winter air temperatures in 2013 were sometimes at -25 F

or colder, with wind-chill factors well below that. For each lake trek, we hiked/skied out to and/or around Crownhart Island in the center of the lake on 119 individual, different days total...the most we have ever "ice hiked" in one calendar year.

We always gear-up thoroughly, remembering my old National Ski Patrol and Eagle Scout adage: "There is no such thing as cold. There is only lack of proper preparation." What great and often challenging days they all were. We were always well prepared for each adventure.

Then, Mother Nature gave us a new challenge. In spring of 2013, despite our thorough preparations to celebrate a special milestone family birthday outside on April 27, our plans were altered by lasting winter days. The dense snow cover in our front yard was approximately 27 inches deep, matching the date, so we switched Debbi's decadal, outdoor potluck picnic/campfire party to a local indoor venue: a willing hardware store. Our last brief backcountry skiing trip on the trails and lingering lake ice in 2013 was early morning on Thursday, May 7. An awesome, powerful winter to remember.

Just wait until we tell you about our cold-weather lake adventures in 2014. And did I mention my 107 ski days in 2012? This is what small-town fun is like, here in our own snow-globe winter paradise.

Chapter 29

Punkin's Christmas Snacks

Our nine-week-old Sheltie-mix puppy, "Punkin," orangey-brown like her name, was the size of a rabbit. We built her a plywood box to spend nights in, safe and secure, at our little, brand-new home. With our new carpets and furniture, our first Christmas was in progress. Short on funds, yet eager to celebrate in splendor, we baked salt-ceramic ornaments for our Christmas tree, decorating with lights and handmade art. We were pleased to have a beautiful tree decorated in a unique manner on our limited budget.

On that first night our tree was up with our new puppy living with us, we nestled Punkin into her kennel box, and went to bed, closing our bedroom door. She made noise early in the evening, then was quiet. "Great puppy!" was our conversation in our bedroom. We slept through the night fine, then got up the next morning to find our Christmas Day surprise. Punkin had enjoyed midnight snacks.

While we were sleeping, our little canine housemate had gotten out of her box somehow and explored the house. Those colorful, so-clever, salt-ceramic ornaments we carefully made and hung on our tree seemed absent. All the lower ones were missing, and Punkin was just lying on the tile floor in the kitchen. She acted sick. Our brand-new carpet, just installed a few months earlier, looked OK to us at first, but then we noticed an unusual background odor, competing with our other seasonal house smells of Christmas pine, baking and spices.

Punkin had indeed explored her new home. She also ate most of the lower homemade ornaments we had carefully placed on our tree. The combination of flour dough, salt and our hand smell on the baked decorations drew her attention. To her, these were tasty treats. She had swallowed many, then so discrete, threw them up and pooped in secluded corners of our living room and dining areas. She concealed her deposits behind furniture, shelves and couches. Our pristine wall-to-wall carpet, and our entire new home had a distinctly used aroma.

Punkin recovered fast. She was friendly and content, and of course not aware that she had done anything other than the sample delights we had made and placed out, just for *her* enjoyment overnight. We took her for a long walk, pottied her around the yard, then brought her back inside for a Christmas morning nap on the kitchen tile floor.

We got out our buckets, disinfectants, brushes and well-worn vacuum cleaner, then proceeded to scrub up remnants of Punkin's solo Christmas Eve revelry. Our new home was not as new as the day before because we tried to save money by making our own ornaments. Punkin was fine and unaffected. Her focused parents stocked up on pet cleaning supplies and tweaked her little kennel box to prevent future overnight house explorations. Despite our miserly tendencies, we bought new, non-tasty ornaments for future house trees.

Never fully removed, Punkin's puppy spots became reminders

of mutual human and dog maturation in our cozy, once-so-new home. Punkin instructed us about including other critters into our home, and maybe, someday, even new, tiny people. Punkin's midnight snacks perturbed our pursuit of perfection, encouraging us to begin thinking about additional family...but never again salt-ceramic bobbles.

Chapter 30

Preschool Drug And Sundries Pusher

That was me. At ages three, four and five, in the early 1950s, I helped out as best I could in our family drug store: Kings' corner pharmacy in Elkhorn, WI, population about 2,000 at that time. We were the only pharmacy in our bustling little town and rural agricultural area.

My Dad, Victor, the registered pharmacist, and my Mom, Madeline, the degreed elementary-grades teacher turned front-of-store manager, showed me where on the hardwood floors to slide the dusty shipping boxes of aspirin, chewing gum, cough drops, toilet paper, cold cream and Kotex. We also moved cough syrup, toothpaste, Kleenex, mineral oil, candy, and other drugs, cosmetics, candies and sundries we sold.

The large brown boxes off the trucks were tattered, dusty shipping cardboard. I usually could not lift them. But once the adults put them on the store's smooth hardwood floor, I could scoot them, one at a time, into place amid our rows of retail shelves. Yes, I became a young drug and sundries pusher. In fact, I got fairly good at it by using my legs, arms, back and butt. Based on my folks telling me this years later, I first learned to read from my box-pushing chores. I would match the sign or tag on the store shelf with the printing on the shipping box, usually getting each carton moved into about the right position on the front, shelf-crowded, box covered floor. It was working fun.

Dad or Mom, or one of their clerks, would come along with a utility knife, and slit open the outer cardboard of the boxes, then show me just where to stock the product. I learned to carefully insert each smaller item, whether bottle, can, roll, bag, or little box, onto its shelf home so that it matched and lined up with the others of its type.

Positioning Kleenex boxes exactly so the words all matched and faced the right way was my special job. I was entrusted to do it alone. And I usually got it all in proper order, according to what my folks told me years later. Printed box lettering was so interesting. During the busy Christmas and winter holiday weeks, and throughout the year, I often stayed at the drugstore with my parents for their evening shift until their 10 o'clock pm closing. Box moving was fun but tiring.

My special hiding hutch in the back of Vic's prescription room was a large empty Kleenex shipping carton: a big brown cardboard box: my pretend cabin. I could play, listen to the busy chatter of our retail world around me, hear stories on Dad's counter radio, and snooze.

My parents put a couple of blankets into my box, plus my Teddy bear, who had a name that I can't remember. That became my quiet, cozy place to escape, play, and fall asleep, especially when they were working into the night with late customers and inventories.

On Christmas Eve, we closed the store and went home at 5 pm to have dinner and exchange family presents. By 7 pm, the phone calls started at our home. Dad did all the deliveries. *Ring. Ring.*

"Vic, I need a refill on heart medication. Bring it over right away."

"Sure, Elmer. I'll be there in a half hour." Dad was always polite and prompt with our customers, even though it was Christmas Eve.

"And put in a half gallon of chocolate ice cream and a box of Whitman Samplers, too. OK, Vic?"

"Yup. That's fine..." Dad was organizing his thoughts and route.

"A pint of peppermint schnapps would be good, too, Vic. Put it on our account. Merry Christmas to Maddy and the kids."

The calls kept coming. That was our usual family Christmas evening, then the daily grind of long hours, 7 am to 10 pm, resumed on Christmas Day. New Year's Day and Easter. All holidays were similar for us. Folks needed their medications, and their other essential treats and sundries.

Throughout the year, our Bastian Blessing soda fountain was also a constant draw for customers. Our soda-fountain-counter workers were busy from early morning until 10 pm closing because people mainly drank Coca-Cola at that time, rather than coffee. I can still see the dozens of soda-straw casings sticking to our tin ceiling above the fountain area. High schoolers and kids put their soft, chewed gum onto the top end of the straw wrapper, then blew it upward to stick on our high ceiling. The tin ceiling over the soda fountain usually looked like it had thin, tubular white stalactites hanging down. Hmm. A *real* cherry-vanilla ice cream soda would taste great, and a hot-fudge sundae in a conical paper insert in the chrome holder... Just a thought.

No more boxes to push. No shelves to stock. My drug and sundry days taught me about words, numbers, spelling, and order. As I recall the feel, sounds, smells and tastes of my drugstore years, they remind me that necessary work must get done. Work can be a fun, enjoyable challenge, a game and not a chore, if I let myself see it that way. Pushing drugs and sundries around our store floor was a good start.

Chapter 31

Shoes Worth A Fight

Hank's only shoes had disappeared from his gym locker. As a kid from a poor family, Hank, a new freshman in 1926, emerged as a quiet, durable, running back on the high school football team. One day, after an early-season practice, he was told by other team members that a varsity lineman had taken Hank's only shoes from his locker. Leather shoes were rare, treasured commodities in those days for his hard-luck farm family. Angry and undaunted, scrappy Hank got them back.

After his father's death when Hank was eight, he and his mother continued to work their farm alone, but eventually had to move into the nearest small town, where Hank would begin ninth grade. His move from the country school set him back a grade, so he was a year older than many of his classmates.

As a small, but strong, alert farm kid, Hank was soon recognized as a canny, fit athlete: a smart, cool-headed player whom others admired. Teachers and coaches urged him to join the high school football team as a ninth grader. He tried out and was recruited as a running halfback for his speed, agility and grit.

At this time, about 1924, Hank and his mother, Lena, were

nearly destitute, with few possessions, personal goods and luxuries. Hank was still wearing the one pair of his father's leather shoes left after his death a decade before. Hank had grown into his father's remaining pair of shoes, still too big, but they were Hank's only footwear for school.

Early in his freshman football season, Hank came back to his gym locker after a practice to find his only shoes were gone. Embarrassed and riled, he would have to walk barefoot for the mile or so home. But first, he had to walk several blocks from his high school to the public library downtown. He was meeting up with his new sweetheart, a town girl, at the library right after football practice that afternoon.

As Hank walked barefoot on the sidewalks, nearing the library downtown, he spotted Dutch Dougherty, an older, bigger varsity lineman on the football team. Dutch was walking on the other side of the street, flaunting Hank's only shoes: leather shoes that were among his few lasting reminders of his deceased father. He missed his dad a lot. At 5 feet 6 inches, 150 pounds, Hank was farm-strong and *angry*. He rushed much-bigger, cocky Dutch across the street, hip-tackling him and knocking him down on the pavement. Then Hank took back his shoes, pulling them off Dutch who insisted he just borrowed them. Hank tied his reclaimed footwear tight onto his own feet.

Hank's courage saved his pride, dignity and reputation that afternoon. He became known as the determined, durable, no-nonsense new kid in school, earning the respect of classmates and community. And he got to his date with his new girlfriend, winning her admiration, and beginning a lasting love that spanned nine decades of long, active lifetimes together, both living into their nineties. They were married for sixty-five years and raised two kids. One of those kids was me.

Chapter 32

Treasured Golden Elephant

Our older son was four years old when we went on an adventure to our local bank in northwestern Wisconsin. In our little town, we had rented a new safe-deposit box at the bank for important things we needed to protect for our young family. Baby photographs, home records, appliance receipts, car documents, birth certificates, precious pictures of our ancestors, and so many other fragile items were essential to preserve. Working together with both of our small sons, we chose and put items into a shoe box at home to take to the bank. Adam and little brother, Seth, were both intent on the task, observing how and what we packed for our walk to the bank the next day.

After our short hike, with our boys and us taking turns carrying our box of memories, the helpful bank clerk took us into a big open, vault where we signed time-record cards. We handed her our key, and along with her keys, she took out our little locked box in the stacks. She handed it to us, then led us to a small private room so we could put in the treasures we had brought.

Our boys watched, fascinated, as we carefully listed and put in each of the documents and special things we selected from home.

It became quite a family adventure for us, addressing that central question in all our lives: *What are the material things and memories we truly value?*

Just as we were about to close the box to put it back into its shelf, Adam handed us the tiny, gleaming-gold ceramic elephant he had carried in his pocket. Santa, or one of his helpers, had placed this little treasure in Adam's stocking during our recent Christmas holiday.

"Can I put this in, too?" Adam asked. We wrapped his little elephant in a tissue from Mom's purse and gave it a safe new home in the solid metal box.

So many challenges and life changes happened over the next, many years. My sister and our parents passed on, we moved to new homes in new towns, plus family farms, homes and properties were sold. Our birth certificates, insurance papers, old driver licenses, and the inevitable death certificates, plus myriad other evidence we were alive and active in a document-directed society, all accumulated in our safe deposit box. Someday, we must go through it and sort everything out.

Throughout fast-rolling years, Adam's elephant remains guardian of our secure things, a role to which Adam wisely appointed him long ago. The tiny toy watches the changes of our lives. He is dependable, always there, along with our memories of little Adam on that day.

Both our sons have moved on to wonderful careers, families and homes, now with their own important items to secure. We think of them often in their distant, hurried, responsible lives, marveling at their demanding worlds and knowing that we can never fully enter again.

Next time we go to our safe deposit box, we may get Adam's elephant and send it to him. Maybe he could use his faithful guardian to watch over him and his own family's important items now.

Yet, Adam's treasured elephant remains our dependable constant: a tangible reminder of trust and love continued from innocent, little-boy days, so long ago. Right now, we need Adam's treasured elephant just where he is, our silent sentinel of priceless memories.

Chapter 33

Just One Day

We moved back to Solon Springs in 2004, to our small home in this old-growth forest clearing. We were gone too long. Every day, we work and play near our Red Pump, the center of our lives, at our cherished, pristine homestead, now into its third century and second millennium of our family's stewardship.

Life here has become slower paced for me. At last, I have time and opportunity, after all those years of not truly knowing my origins or family's stories on both sides, to at least begin the involved, detailed process of going through boxes of family pictures, letters, and other memorabilia. It has been a time of diligent re-acquaintance with my parents and older sister, who are now all deceased, and of getting to know my grandparents and great grandparents whom, with one exception, I never met.

My view of life, the world, and of them, has changed. How wonderful to see Mom and Dad, even grandparents and great grandparents, in long-stored pictures as teenagers, and to read some of their goofy personal letters and cards. They were young once, as I was, and seem so familiar.

Sweating in our barn attic, as I sort through dusty boxes, is an adventure I had not imagined prior to our move back here. When I have time, I view and handle items, indeed treasures, some recent, some from long ago. It is much like entering a personal, family time machine. I see pictures and read of births, graduations, new jobs, lost jobs, and of illnesses and deaths, of triumphs and tragic family losses, especially in way too many wars. It is quite a rollercoaster ride each time I look through things, learning more about these people of whom I know so little. They are becoming more real to me.

Some days, I find my exploration to be wonderful and exhilarating. On other days, as I read enthralling letters and documents of challenges, disappointments and sorrows they all endured, through some gracefully long and some tragically short lives, I find it humbling. It is an honor to be the latest in our family succession of those who lived fullness of life here, amidst pristine lakes, streams and forests.

As I look through more things, my thoughts wander. I have come up with an interesting thought experiment. You may wish to try it: Imagine for one day of your choosing, just 24 hours, you, your parents, your grandparents, your great grandparents, and your great greats, as far back as you want (as well as your own children and your siblings, if you wish), can all be together...and that you will *all* be the *same age.*

What age will all of you be? What setting will you be in? What will you say and do? And why, for all of those? This imagined scenario is interesting to work through. We can only wish it could happen, but it has caused me to examine relationships among and to long for interaction with relatives I have never known. It has also caused me to think in a focused way about past family lives and experiences which have allowed my family today and me to be as we are. This exercise can be valuable for all of us history and genealogy fans. You may be surprised, even delighted at what you find about yourself and your family as you try the experiment. Take a moment to think it through.

If you had *just one day*...

Chapter 34

Magic Snow Socks

Abagail darted through the packed, narrow lanes of new January snow. Duck-Duck-Goose was the game. Her kids called her *Sister* Abagail, and they couldn't catch her. The good Sister's secret was her knee-high woolen socks: she wore them on the *outside*, over her other stockings. No boots. No shoes. Sister Abagail's thick, hand-knit socks gripped the dry snow, giving her tremendous (seemingly *divine*) traction advantage over her young pupils. They loved it...and *her*...following their agile teacher in games of Tag, Cut the Pie, Pom Pom Pull Away, Red Rover Come Over, and foot races around the schoolyard.

Sister Abagail, perhaps a decade older than her students, always won. Laughter and squeals of excitement set aglow this hardscrabble valley of northwestern Wisconsin in early 1917.

A timid, dark-haired boy, almost nine years old and always alone, walked past the lively group each day to and from his own rural, one-room public school just down the road. Victor, in fourth grade, shared his school with kids of many ages, and with their teacher, Miss

Picotte. During his school hike each day, Victor noticed the activity at Sister Abagail's playground. Her winter exuberance with the children was palpable, contagious. Several months earlier, Victor and his mother, along with a few relatives, watched as his dad's casket was lowered into a cold, gaping grave. Soon after, Victor's older, *only* brother left for the new war. Play and laughter had become rare in Victor's life.

On a sunny late-January afternoon, Victor finally stopped on his way home to watch Sister Abagail's excited group. A dozen or so boys and girls about his age tried to break the chain of gripped hands in their enthusiastic playground game of "Pom". Some of the kids looked familiar. Soon, the friendly, athletic Sister ran over to him.

"Would you like to play?" With her warm smile, she offered Victor a place in their game. "New snow is fun!"

Self-conscious and timid, Victor looked down at the snow and his rough-shod feet. He was poor.

"No, I can't today." Victor still had on his barn boots. His only other footwear were the old lace-up, leather-soled shoes from his recently deceased father's closet. Those were sad and comical, way too large, too slippery, too worn to walk in, let alone to play in with new kids. Embarrassed, Victor explained that he was wearing ragged, dirty socks from his morning barn work, keeping them on because he rushed to get to school on time. His socks had big holes needing darning, and he didn't want the kids to see.

Sister Abagail said she understood, and that he was welcome to stop by anytime. The sparkle in her eyes beckoned to Victor, along with the inviting gleam of bright snow. They bespoke a promise of fun and inclusion...sometime. He walked home, thinking.

Several days later, as Victor did early-morning farm chores, the new powder snow was over his boots. His rural school seldom closed in winter, except during true blizzards, because the twenty-some kids enrolled all lived on nearby farms.

As durable, farm-strong children, they either hiked to school, rode on their family's horse-drawn sleighs and wagons, or slid quietly over the fields with their farmyard skis. Each winter school day was precious because family farm work often intruded on the kids' attendance during spring and fall months, when their help at home was essential.

Today, Victor would rely on his simple, handmade skis. They had large leather toe straps, allowing him to wear his barn boots, and to move easily among farm outbuildings in deep snow. He slipped the toes of his boots into the straps, and off he went for chores. And that was how he would go to school, skiing across fields over much shorter routes than he could walk on roads. Victor knew his schoolhouse would be warm and welcoming, with a roaring wood fire that Miss Picotte built in the barrel stove. She and the older boys made certain their school was always cozy and safe for the smaller children.

Victor was excited. His mother packed his lunch of two thick slices of homemade bread spread with lard, a small glass bottle she filled with milk from their one remaining cow, and a compact jar of maple syrup from their home sugar bush, for dipping his bread.

When his school let out early that afternoon, Victor skied toward home, staying on snow-covered-field trails that led past Sister Abagail's group. They were already out in the snow, playing and shouting. Even from a field away, Victor could hear the loud game of Tag, as the children and Sister sprinted and turned, twisted and dodged past each other to avoid becoming "It". Laughter filled the valley.

As Victor skied up to the parochial schoolyard, he saw Sister Abagail running toward him in her black-and-white habit, with her distinctive wool outer socks gripping the sparkling snow. She was agile, quick and sure-footed, and holding something in her hand. His heart soared.

"Hello, Victor! Would you like to play with us?" She *knew* his name.

Victor nodded. "Yes, I would. Thank you."

"Very well. Just place your skis and boots over there by the tree. Pull up your socks as tight as you can and tuck your pants in them. Then pull these big socks over the top of everything."

Sister Abagail handed him a new pair of bulky, gray socks she had knitted. Constructed from coarse, left-over rug yarn, they were rough, warm, and just what Vic needed to join the fun.

"They're yours, Victor." She spoke so just *he* could hear.

Victor put on the socks and tried them out; marveling at how he could leap, run, stop, and turn so fast. These were the best things he had ever worn on his feet. He joined with the kids and Abagail as they lined up on two sides of the playground for Red Rover. Their second call was "Red Rover, Red Rover. Let Victor Come Over!" Victor darted and dodged to the other side. In free! The bulky snow socks worked, again and again. *Magic!* So nimble. Such grip. Victor enjoyed new warmth of acceptance, skill and admiration.

For an hour that day and on days to come, Victor found haven from sadness and poverty, an affirming refuge he could share; so different from what he had known over the past year. His father's death, his only brother's departure for war, and farm responsibilities, now Victor's and his mother's alone, all faded in the glistening snow-globe world Sister Abagail and her students extended to him. Victor stopped by often to play during that exceptional winter of Magic Snow Socks.

Note: Later in his life, Victor, my father, told and taught me about the magic of thick wool socks on cold, dry snow. He spoke often of the miracle of comfort and secure companionship he found with those kids, and of course, the warmth of dear Sister Abagail, his welcoming, rescuing angel. Victor believed she was truly heaven sent, just when he needed her. Sister Abagail's love of snow and fun and him, changed Victor's life.

Chapter 35

Spartacus, Our Durable Squirrel

Speaking of dedicated nerds, all around the globe we amateur radio operators love to talk, make new friends, and engage in public service as we also conduct experimentation and research. Those passionate pursuits are in our blood, regardless of our nation or language.

So, here I go... "CQ CQ CQ... WF9I, *Whiskey Foxtrot Niner India,* calling CQ twenty meters and standing by." My general CQ (*Seek You*) call for radio contact goes out to the planet this autumn day in 2016, just after sunrise. I transmit on 14.271 Megahertz, upper sideband, hoping a distant amateur radio station will hear. Again. "CQ CQ CQ..." No response. I turn back to my writing.

Thirty minutes later, I try once more, using my best, strong Ham radio voice. "CQ CQ..." in clear cadence. No answer. Back to work.

I type a few more lines. Sudden motion, outside to my left, sparks my attention. Something is on the deck. I glance through our large windows overlooking old-growth pine forest and nearby lake in Solon Springs, Wisconsin. *Eyes* are on me. I feel them, turning in my chair to look.

A large Black Squirrel, on the closest deck rail, mugs at me, staring through the window, near my ladder-line antenna feed, trying to catch my gaze. Torn, bitten, bloodied...his coal-black pelt is ripped from head, neck and abdomen. Large, oozing crimson bite punctures gape throughout his shiny ebony fur.

Dark eyes, close against the glass, plead "Help me!" Gaze and posture are poignant, irresistible. His full-Monty stance facing me confirms he is a boy; a courageous guy who just had a life-and-death tussle with a coyote or wolf, maybe a fox or fisher in our shadowed boreal forest. In this northwestern corner of Wisconsin, wild predators and prey are part of life. Black Squirrels are common variations of our abundant Grays. Their dark color provides warmth and camouflage advantages: bushy tales serve as umbrellas, parkas and blankets.

This inquisitive fellow heard my rhythmic CQ call for contacts. He is seeking help. *My* help. I speak to him softly through the huge window. "*Spartacus*... Yes, that's you." My tough survivor-visitor, this ripped-up rodent, needs assistance. A bag of in-the-shell roasted peanuts is a quick sprint away in our kitchen. I grab a handful, flinging them on the deck.

Spartacus doesn't flinch. He watches from the rail, then leaps down to the deck boards, trying to snatch two peanuts with his crushed, bleeding mouth and jumbled, broken teeth. This brave dude fits in only one, then scoots back up on the rail to eat it. I notice a nearby *drey*, a squirrel nest. We look at each other, curious, not an arm's length apart.

Squirrelly gratitude is evident. I toss out a dozen more nuts, then close the door, returning to my desk. Immersed in my work, typing and concentrating hard in my flow state, I lose track of all else around me.

In late morning, I look up, thinking of our new neighbor, hoping he is doing better. No one is on the deck. I still have that itch for an on-air chat, so using my most-resonant radio voice, I again call "CQ CQ CQ... WF9I..." No luck. Using Morse code (Hams say "CW"), I send CQ as rhythmic *dits* and *dahs,* waiting. Nope. As I return to work, I sense movements on the deck.

Sparty (we had become quite informal now) is back. He bounds up the rail near my window, looking directly at me with plaintive, appealing indigo eyes. His wounds seem better. I see less scarlet on him as I throw more peanuts on the deck. He straddles my RG58 coax lead near its wall entry, then scurries down and up, many times, onto the ground to bury the treasures with his paws. Captivated, I watch, then go back to work. A half-hour later, I look up from my writing for him. No squirrel.

In an hour, I break for lunch, then try calling again, using forty-meter frequencies this time. "CQ CQ CQ..." Contact! A new Ham in British Columbia answers my spoken *Seek-You* call, and we have a fine, long chat, a "rag chew" in Ham-speak. Debbi, N9GLG, my wife ("YL") looks in. Smiling. Thumbs up. Then I shut down my rig, and return to afternoon work, focusing, becoming wonderfully lost in my world of words, forgetting for the time about Spartacus and all else except finishing my writing project.

In midafternoon, jolted from my wordy flow state by deck activity, I glance out our windows. Just across the glass are those dark eyes...that expressive furry face emoting at me with front paws held in close, cute. Durable Squirrel is an effective beggar.

"Please Sir, I want more..." Oliver Twist performed convincingly by this bold, convalescing Black Squirrel. My lifelong Dickens fascination revels as Spartacus sits up, focusing directly on me. His sincere, humble eyes, mouth and ears are persuasive tools. He continues his prayerful pose, with both front paws held close together, just below his chin.

"I beseech you, Sir. More please." That's what I interpret from this affable spokes-rodent. Sparty's posture and demeanor radiate intensity. Yes, crazy squirrel person, me. If Kevin Costner ever

casts for the role of "Talks with Squirrels" in one of his movies, I should audition for it, hooked as I am on conversation with durable Mr. Personality, Spartacus.

More peanuts? Of course. I flip a large handful out on the deck, then go back to my radios. I watch him as I call CQ on air a few more times, while he watches me. Then I close down for the day to do other tasks around the house and in our little village in the forest.

Next morning, I am back at my radio desk, soon after sunrise, writing near the windows. As an experiment, I loudly call "CQ Squirrel, CQ Squirrel" while I step outside on the deck, spreading a cup of fresh peanuts and sunflower seeds on the waist-high rails. Sure enough, in moments, appearing high up in a nearby maple, on a limb near the north tie-off point of my 80-meter dipole antenna, a small dark face peeks from behind bright orange leaves. Spartacus scampers up to the deck. More peanuts. I go inside to write. Sparty eats. My peanut bag empties. Gradually, he is recovering and looking less tattered, with wounds drying and shrinking. Squirrel and this easily manipulated radioman/writer bond more each day.

My manuscript draws me back into its world, but as I try to concentrate, a large, fuzzy Gray Squirrel, with blonde ear tips, hops onto the deck for treats. A female. She and Sparty assess each other, pause for a few moments, then posture, circle and tussle with delighted abandon. They chase around and around, then up the nearest towering Norway pine, where the apex of my multi-band "sloper" aerial is mounted. Frenzied pursuit. No eating. Little OM (Old Man) Sparty is feeling better. Durable Squirrel has found his fine Young Lady (YL).

Rodent romance is in progress. This story is getting interesting.

MONTHS PASS. Throughout our deep winter of 2016, and into spring of 2017, Sparty and Bright Ears are on our deck each day, joined by many, including Grayson, a chunky Gray male; Bob, a small Black Squirrel with half a tail (Get it...*Bob?*); Stumpy, with no tail; and Bullseye, a young female Gray with a healed bullet wound on her left side. All our squirrely-folk friends fattened up well during fall. They coast now on plump internal body stores, along with what they glean from purchased squirrel mix we put out for them daily, and from edibles they buried earlier in our yard and forest. Temperature or snow usually do not affect their arrival on our deck each morning. As the sun gets higher and brighter in the sky, and as temperatures increase, they stop by most days.

By late March, we see new adolescent squirrels on our deck, fully furred in an array of colors, about three-quarters the size of their parents. "Junior High" squirrels, we say. These kids, now out of their snug, leafy drey or tree-hole den, nearly independent, work along with Mom and Dad, following their tree-top highway to tree trunk to ground to deck. We marvel at family patterns of behavior and colorations, imagining their song, rendered in tiny, enthusiastic squirrel voices, as they troupe in each day: *"Hi Ho! Hi Ho! We're hopping through the snow..."*

Eastern Gray Squirrels (*Sciurus carolinensis*, "shadow tail") display melanistic fur variations, genetic expressions of three basic colors: orange-brown, white, and black. One in ten thousand individuals is a Black Squirrel. Our Spartacus-Bright Ears offspring show a palate of squirrely hues. Each day, our rodent theatrical cast play and feast on our deck. Fifty shades of Gray Squirrel celebrate each other. Frisky, tumbling squirrels party several times daily, right outside our window.

Amazed and amused, we watch with joy. There is always a new floorshow with each appearance of the family as they chase, cavort, nose bump, and pull rank on each other for treats.

Outside, I adjust my wind-worn antennas in our forest, and

mow lawn into the hot summer, watched by squirrels above me. The growing Spartacus family have built many cozy, aerial *flets* about thirty feet above ground. These sprawling, leafy nests are in tall oak, maple, and poplar trees throughout our five-acre forest, often anchor trees for antenna support lines.

Squirrel Harbor here. Kings' Home for Wayward Rodents.

Spartacus, my active little Old Man, now a grandfather squirrel, watches this old grandfather from nearby trees as I work outside. He sits on branches above me, staring, rubbing his front paws together, always hungry. Bright Ears and their white-mittened kids watch, too, but never beg with their father's intensity. Sparty is bonded to me, even looking in through house windows as I move about inside. I shave in our upstairs bathroom most mornings, and sure enough, there he is, with little eyes and beseeching clasped paws watching from a limb on the oak just outside my window. Imprinted to my voice, Durable Squirrel shows up whenever he hears me calling CQ on our radios, or "CQ Squirrel" from the deck. Yes, we are good friends.

Each contact I have with distant amateur radio operators now includes discussion about Spartacus and his furry clan, my silent on-air partners. Some days, Sparty peeks into my open, outside door, but goes no farther. My numerous confirmed radio contacts, each known as a "QSL" in Ham radio parlance, are exceeded by my face-to-face QSLs with Spartacus. We interact with each other several times daily. Beeps of Morse code, along with other short-wave noises from my radios, let him know I am there, ready and persuadable to interact. My voice and other unusual sounds alert and summon him, but harsher bursts of weather or solar noise on my sensitive receivers frighten him. He looks into my radio room, staying outside. Courteous and cautious, he knows this is my indoor world, not his.

One early evening in midsummer, we watch from our living room as adult and junior squirrels scurry and rustle in thick, nut-laden hazel bushes in our forest. Debbi and I laugh, pointing, hooting as we watch them jostle, fall and romp, with wild acrobatics of posturing sideways, then upside down for the tasty nuts. In an instant, a Red Fox pokes her head out through the bushes, grabs a Black Squirrel, shaking it hard, side to side, breaking its neck so it can't scratch her. Blood squirts and splatters as the squirrel screams, then goes limp. Momma Fox clasps her ebony prize, with her powerful jaws around its middle. The motionless squirrel hangs limp, dying in moments. Foxy struts across our yard, head and tail held high, to feed her kits at a den farther into the forest.

Debbi and I gasp "NO...not Sparty?" We aren't certain, but our fears mount. We know this is the natural course of forest prey and predators, yet so hope it wasn't our friend. We wait.

Next morning about 8 AM, there are Bright Ears and many kids...and Durable Squirrel, together on the deck. Whew, not Sparty's time. We are surprisingly emotional about it all, relieved and grateful beyond measure. Our favorite little guy is still with us.

From early 2018, through our cycle of seasons into spring 2019, Black and Gray Squirrel numbers grow here at Sunny Cove, in our bright, lush forest near our home and Ham station. Spartacus and Bright Ears, with many more kids and grandkids, populate our deck and yard. We notice new, high-up flets that maturing Sparty kin have built in trees around our yard, way into the deeper forest. Curious, we read online that Black Squirrels live about five years in the wild. We ponder how we first met Sparty, still so active, when he was about age two. We realize what lies ahead, as we savor each day with him and his crew.

During winter into mid-April of 2019, we enjoy Sparty's daily visits and those of his many family members. Then that dreaded day beckons.

We see the sleek, beautiful Red Fox crossing our woods road, carrying a limp, older Black Squirrel in her mouth to feed her family. Our hearts sink. We gulp, having witnessed this before, then being relieved to know Spartacus lived. On the next days, we make a constant watch for our Durable Squirrel, our tenacious survivor, to reappear.

Spartacus does not return. His family romps on the deck each day. Some respond to my "CQ Squirrel" call. But my heart is broken. I miss my squirrelly buddy, who brought us so much joy and connection with our natural world here at Sunny Cove. We will miss him, fondly recalling how he followed me around home and yard, watching me with those appealing dark eyes as he reliably showed up near my antennas and deck, ready to interact and make me spring to action for his treats. Sparty had us so well trained to give him food and attention. He will always be among our most memorable boreal-forest acquaintances.

We helped sustain his life. He enriched ours.

Rest in peace, dear Old Man. You gave us so many fun, intense face-to-face contacts, eyeball "QSLs" as Ham enthusiasts say. Here at Sunny Cove, N9GLG and WF9I, Debbi and I, amateur radio operators accustomed to our globally shared shortwave shorthand, know that, for us, our three-letter Ham abbreviation **QSL** will forever more stand for our three treasured years of excellent inter-species contacts with our singular **Q**uality **S**quirre**L**: Spartacus, our dear, durable forest friend.

Part Two

SCIENCE & TECH

Chapter 36

Nerds In Universal Motion

A mark of true nerdom is to consider ideas and concepts that may not be apparent to others, and then to pleasantly obsess on them into minute, detailed precision. We, who delight in savoring all types of information and complex concepts, find great comfort in pursuing an idea to extensive depth, even to exhaustion of ourselves and the topic.

So, as I was eating lunch, and strategizing about errands I needed to do for the day, it occurred to me that we lifetime learners, and all others who move about on our exceptional planet, are also in constant motion through the universe, even when we think we are just sitting still.

My glasses, phone and keys are on the table next to me. I set them down a few minutes ago. And they are still right here, relative to me. Yet they have already moved far beyond what I can comprehend. I had to look this up and get the specifics of exactly how we are moving, even when we think we are being sedentary. As you might well anticipate, I found more than I had expected.

Here is my attempt at a quick, readable summary:

At the equator, our Earth rotates at about 1,000 miles per

hour. That's why rocket launches to space are done nearer the equator than where we live in mid-continent North America. Our planet acts as a spinning catapult. Here in Wisconsin, we move at about 700 miles per hour, as our planet turns on its axis. We may be aware of the movement in tides, night and day, and seismic changes we sense as earthquakes and temblors. Our planet responds in many ways to the complex physics of its movements.

While we spin, Earth simultaneously orbits our star, the Sun, at about 66,000 miles per hour. We seldom notice this motion, but can observe our circling around our main star by the changing constellations and star patterns we see throughout the year, a full trip around our Sun.

Adding complexity, our planet follows the Sun, as our entire solar system orbits as a group within the Milky Way Galaxy. Our collective movement is around 43,000 miles per hour, as we travel the circuit together, our Sun with all the planets in tow.

Extending the awe factor, the Milky Way, our home galaxy, is traveling through the universe toward constellations Leo and Virgo at around 1.3 million miles per hour relative to reference points in the universe outside of the Milky Way.

To recap: Earth at the equator spins at 1,000 miles per hour, as we orbit the Sun at 66,000 miles per hour, as our entire solar system circles within our galaxy as 43,000 miles per hour, as our galaxy speeds though our universe at approximately 1.3 million miles per hour.

All of this is happening at once. We can measure each variable.

No wonder I have such trouble finding my glasses, phone and keys. Thanks to gravity, they did not fly off with all that angular momentum from rotation. Yet at each moment, they are never exactly in the same place I originally put them. So, where is *here*? And we are... *Where?*

Chapter 37

Speaking From The Ocean Floor

*Just breathe. No technology required.
Speech and Voice. Overlooked gifts.*

Talking and listening seem so simple here on Earth, as we speak from the floor of our planetary ocean of air. We live at the bottom of a protective atmospheric sea, about twenty miles deep. We breathe it to live, and rely on it to speak and hear. On Earth's surface, the piled-high weight of gas molecules (mostly nitrogen and oxygen), pushes air into our lungs at just under 15 pounds pressure per square inch when we inhale. With *in*-breaths, we can trap air below our vocal folds, our airway pillows of muscle that we squeeze together to pressurize exhaled air for voice production.

Our vocal folds (not *cords* or *chords*) function as linked mechanical oscillators. They are protective, "kitchen double-doors" composed of sensitive tissue in our larynx, opening and closing to help us swallow and cough safely, and to voice and speak. Our vocal folds reside just behind our front-of-neck airway bump, our Adam's Apple. For voice production, they act as linked tracheal resistors (think *Ohm's Law* for fluids), oscillators that open and close precisely to capture, impound, then efficiently valve and pulse pressurized *out*-flowing air for speech.

After we inhale, we contract thoracic and abdominal muscles to increase air pressure below our closed vocal folds. Then we exhale to produce phonation--our sustained vocal *buzzzz*...turning our voicing off or on by opening and closing our vocal folds as needed for speech, breathing and airway protection. Air pulses rush out at about 100-200 per second through our oscillating folds during voicing, creating our own personal fundamental vocal frequency. (*Hold out Aaaahh...*) About 110 Hertz or cycles-per-second for males. Approximately double that for adult females.

Air molecules knock together in these sound-pressure-wave pulses, as they billiard-bounce outward, forming our laryngeal source tone, plus harmonic overtones. Our voice and multiple, modifying resonances exit our head, shaped by our own skilled adjustments in our throat, mouth and nose. We learn these subtle yet complex tweaks by listening to other skilled speakers of human languages.

Respiration, phonation and resonation for speech begin with just breathing in from our surrounding air ocean, then individual characteristics of physical speech structures and movements further form our own unique, individual sound (our *per sona*). Our personal vocal sound will change in varying gas pressures or densities such as in planes, submarines, spacecraft, diving suits, partial vacuums. Or, yes, when inhaling from a helium balloon.

Varying elasticity and resonatory properties of gases billiard the vibrating molecules differently than in our customary surface air, making us sound unusual. Also, acoustics of speech change with electronic compression and transmission of spoken words on telephones, radios or in air vacuums. And remember: We can't talk via direct air-molecule connections in space or on the Moon, and not much on Mars where the atmosphere is too thin.

As we produce full voicing, or even a soft *whisper*, our resonated laryngeal source tones are further shaped into individual speech sounds by quick, agile movements of our articulators: tongue, teeth, lips, jaws, cheeks, plus soft and hard palates. Speech articulation in human languages involves fast, precise, simultaneous coordination of a hundred or more muscles, along with a multiplicity of complex nerve firings to accurately produce each speech sound in rapid succession.

Spoken Standard American English employs more than forty of these allowable, diverse phonemes we might also call speech sounds, or even "letters." Other languages may have more or fewer identifiable phonemes, plus tonal variations to each speech sound, as in Mandarin or Hmong. These quick speech-sound bursts travel in pressure waves through atmosphere to the transducing ears, then auditory cortex, of listeners, who may return spoken transmissions...and around goes the cycle of human speech and hearing.

Whether talking or listening, we take it all for granted here at the bottom of our pressurized air ocean. So reliable. So simple. So complex.

Humans have made many clever adaptations to speaking and hearing. We are sharing some right now. Our species wants spoken messages to get through, and to last; even when talking will not work. So we have developed systems of special marks and symbols which we place on stone, wood, paper and electronic screens to represent and preserve what we did say, or could have said with our mouths. We can even just *think* letters, sounds, and words silently to read and write–with no speaking or hearing needed. Written characters allow us to "say" and "hear" within our thoughts, the words and ideas we or someone else marked down just now or long ago. Listening, speaking, reading and writing, along with the thinking, awareness and strategies for use that surround each, are the elements of literacy. They allow our thoughts and words to transcend time and place.

Speech is fast. But writing lasts.

And it all starts by breathing in and out this mix of planetary-surface gases we call *Air*. The physics of our mostly invisible atmosphere, rushing to fill low-pressure systems in our expanded lungs, begins the miracle. It fuels the foundations of literacy, enabling speech and hearing, and all the derivative forms and processes of complex communication we assemble and deploy in our diverse rules and customs of human languages, whether spoken or written.

The terms we all use, *hearing, speaking, reading, writing*, are fluid, iridescent and chameleonic. We interchange them freely because it begins so viscerally, so intimately, as we just breathe in, to speak from our atmospheric ocean floor.

Chapter 38

Trail Trees Are Forest Magic

Trail trees are distinctive. You will likely know one if you see it. And you probably have seen one if you hike in the Great Lakes region of North America. A trail tree is a curious living zigzag: a purposely human-made signal post. It looks like you could climb up and sit on the horizontal perch area.

Sometimes, practical archeology is right in front of us. Spotting a trail tree, then understanding how it and why it came to be, is a unique, transformative thrill. The focus, patience and knowledge of folks from the past who created them can teach us a lot, leaving us humbled.

Indigenous peoples formed trail trees by gradually bending and shaping the trunks of small deciduous saplings into a sharp-angled, left-right bend, forming a flat or nearly horizontal portion of trunk easily visible above brush or deep snow, even at a distance. Other sharp tree bends can occur, too, by natural means, but trees shaped by humans are distinguishable by their smooth and unbroken, bent bark and wood, not caused by wind damage or another tree falling on it.

"As the twig is bent, so grows the tree." is literally true. Trail trees grow larger and become respected, lasting markers of boundaries, trail exchanges, hunting and fishing areas, and directional switches for trail walkers. Some trail trees are large and several hundred years old, and they present their bold prominence on well-used trail locations.

Here in our forest, I have tried bending small saplings over the years to create trail trees, but they have never grown well. Clearly, there is a lost (or at least non-obvious) folk art to shaping a trail tree that thrives and grows into a mature tree. The Trail Tree methodology and art remain undiscovered by me, and but I will keep experimenting.

Yet, other smart, skilled people in many places over millennia perfected it. Be sure to stay alert when hiking trails near Lake Superior and Lake Michigan. You may be honored by an encounter with an ancient trail tree, especially formed years ago by other intelligent, innovative trail users to help guide foot and sled travel on the very paths you are hiking today.

Trail Trees are enduring, memorable delights of the forest. I hope you get to meet some, and consider how, why and by whom they exist. When you stand next to a Trail Tree, and understand its history, purpose and formation, you will sense that it is truly forest magic.

Chapter 39

Tameshiwari: Testing Yourself

Sometimes, our best motivation for learning and accomplishing things we need or desire to learn comes from within, not from external forces. Across cultures, throughout the ages, humans have developed ways to refine and evaluate their own personal achievements.

In traditional Okinawan karate-do, the term *tameshiwari* refers to performing a testing of self: attempting to do something difficult to see if you can do it well. Remember, *kara* means empty, and *te* means hand, so karate is the weaponless Okinawan folk art for empty-hand, weaponless self-defense. The two questions I am frequently asked over my half-century of karate-do training are: How many boards can you break? Have you ever used your karate?

Let's handle that first one: Over their long years of training, students in this art develop, among other skills and attitudes, three first-time, every-time upper body (hand, elbow) techniques and three lower body (foot, knee) techniques that can dependably break through two 12 x 12 inch, ¾ inch thick pine boards, along the grain. These two-board breaks are the approximate equivalent of breaking through ribs, joints or other targets, and require about 600 pounds of quick, snapping impulse force.

Regarding the second question: Yes, every day of my life has involved use of some aspect of karate since fall of 1971. But not as you may think. I have used my art of Okinawan Goju-ryu in self-defense only a few quick times in ways you may not have seen in movies. The difference being that my martial art and my own emphases on fast, humane, simple preventions, escapes, balance, agility, and defenses, are all effective, but are not movie footage because they are too fast, simple and subtle.

More realistically, I have used the most important thing I learned in karate through tens of thousands of knuckle pushups, sit ups, thousands of kata and rounds of sparring with one or more opponents in our dojo classrooms, namely: Tameshiwari, the Okinawan concept best translated as "Self-Testing." In other words, it means permitting yourself to try, do and maybe fail, then eventually, with more practice and hard work, succeed at something you didn't know you could do.

Every time I write, publish, speak, sing or appear in front of a group, whether I am teaching, inspiring, or entertaining, that is tameshiwari for me. Controlling my temper, words and actions is, too. So, yes, I use my karate training, wisdom, strength and endurance every day. And I count on using them a lot more in upcoming self-tests of life and health that I haven't even thought of yet.

Most of these life tests will be pleasant and uplifting, I trust, including writing and finishing this book and doing it well. Writing is precise, demanding tameshiwari at its hardest levels for me many days. But I know I will complete my works. Testing myself, *daily tameshiwari,* keeps me focused, realistic and fully alive.

Chapter 40

Formidable Phillips Code Book
...still with us in so many ways

Ironically, I spent a career as a speech-language pathologist who believed in total communication (all forms: speaking, writing, typing, graphic, symbols, signing, more...) working with patients who lost or never developed oral speech, but who could communicate via adapted methods of writing and typing, and text to speech on digital devices.

The beliefs and professional standards for most of my career were that clients must speak: no other approach was ethical or acceptable. Writing, typing, signing, pointing were not significant substitutes. Some of us were labeled unprofessional for considering non-speech alternatives for patients who could not communicate. Things changed.

At the historical root of our societal change to texting and messaging is telegraphy: sending and receiving messages using Morse Code. And the requirement of the telegrapher to use as few keyer movements (dits and dahs) as possible for maximum efficiency, speed and muscle energy expenditure. Also, telegrams were expensive. The more economical your code transmissions, the cheaper and faster the message could be relayed.

Abbreviations, acronyms, and telegraphic shorthand became a solution. Along came the Phillips Code Book with its listings of keystroke saving letter combinations. We have been there before, and you and I still use the spelling and speaking shortcuts daily.

Electronic instant messaging and text message abbreviations to convey whole messages began with the invention of the binary dit-dah code of Samuel Morse and Alfred Vail in the 1830s. Speed and efficiency of transmission and reception of code, known as keystroke reduction, was enhanced even further in 1879 by journalist Walter P. Phillips, with his book of time-effort-space saving abbreviations *The Phillips Code*. It was published by Telegraph and Telephone Age, 25 Beaver Street, New York 4, NY, and revised in 1945 as the telegraph, cable, radio editions. Interesting and practical. Be sure to check it out.

If you have ever used *LB* for pound, *NO* for number, *CUL* for See you later, *ES* for AND, or *BCNU* for "Be seeing you," then you have tapped into the Phillips Code reservoir. Morse code and the Phillips Code Book shaped our ways of writing, abbreviating, and thinking, and of compressing meaning into fewer keystrokes: exactly what we do when texting. Mr. Phillips, incidentally, went on to become vice president and general manager of United Press news agency.

The "inverted" pyramid journalistic style of putting the most salient facts of a news article into a first topic sentence derived from the fragility and expense of early telegraphy to communicate essential information and facts in descending order of importance and detail before the telegraph lines were cut or went down. Historians of journalism cite the first major use of electronic telegraphic reporting during the American Civil war of 1861-1864. Telegraph operators had to get the most possible information (i.e., keystrokes) into their first sentence or two before lines went out. Inverted-pyramid style still guides our journalistic writing today.

When the Phillips code book came out 1879, and then its

subsequent three or more revisions over the next seventy years or so, it reshaped how telegraphers and writers from all fields jotted things down, including present-day writers of texts, messages, and notes.

You and I, and all who write in English, use these telegraphic abbreviations daily without even thinking about their origins. The next time you type a text message or write a note, remember Walter Phillips and his handy, transformative abbreviation books. So... 73 es C U L

Chapter 41

Modern Morse Code In Rehabilitation And Education

Statement and handout prepared for meeting at Adaptive Design Association NYC in 2014.

Sometimes, a two-hundred-year-old technology, kept alive by focused, visionary folks who retain the art and see the promise, can still be useful. That is the case with modern Morse Code, and related switch codes. Correctly harnessed, these proven adapted-access methods can offer fast, simple and effective, alternate low-tech to high-tech ways to write, type and speak. They can move the world for some users, in conjunction with state-of-the-art digital technologies. Google *Gboard* and Samsung *Good Vibes* are examples of modern Morse Code in high-tech action today.

Complex human communication happens through precise, controlled movements. Whether we push air through our vocal tract and speech articulators, or mark with a pen on paper, or spell out silent equivalents of spoken words with fingers and hands, we make controlled movements to send information. Speech and writing are essential aspects of literacy. Modern Morse code can offer access to other ways to speak and write.

Adapted Modern Morse Code movements can be entered to desk-top, lap-top, or hand-held digital devices by way of special switches, or via standard touch-screen inputs. As with touch typing or handwriting, code-entry movements fast become sub-cognitive, automatic patterns. Morse code is just a set of organized, basic gestures (left-right, up-down, puff-sip, etc.) representing letters, numbers, words, and ideas. Code movements, quickly harnessed by microprocessors, allow users other ways to write, type, and speak via electronic devices, even when spinal cord injury, ALS, blindness, learning difficulties or other user challenges may interfere.

Because patterned movements can be converted to text and speech by common communication devices, Modern Morse code might well be considered the manual language of the new millennium. Morse technologies have existed for some time, but specific focus is needed now on teaching Morse code efficiently and effectively for diverse users and current devices. Solid teaching and learning methods exist, but are often not applied. Emphasis must continue on awareness of the potential of Morse code in literacy, and on fun and efficient ways of Morse code learning by students and clients, as well as their families, teachers, and clinicians. Google *Gboard* offers a free, effective code-teaching app.

The Morse 2000 Worldwide Outreach, which we developed in 1994 at the University of Wisconsin-Eau Claire, and my Modern Morse Code textbook with Allyn and Bacon, Boston, in year 2000, were early steps in disseminating new methods and information. Now, renewed efforts, with state-of-the-art equipment and techniques, are needed to teach others the simple power of this old, proven method. Morse Code for literacy could be availed by more users, who otherwise believe writing or speaking will never be part of their lives. Teaching Modern Morse Code, and how to effectively use it, must be continuing, essential efforts.

Background: This statement about Modern Morse Code was prepared for and sent to Adaptive Design Association, Inc. New York, NY, in 2014, to provide information and ideas for professional staff and families considering alternate ways to write, type, dial, speak, and access computers for persons living with severe disabilities. It helped contribute to Alex Truesdell's wonderful honor in our ADA collaboration, with receiving a MacArthur award, plus our collective liaison with Google in development of the new Google Gboard inclusion of Morse Code.

Chapter 42

Madeline's Minerals

Recently, I bumped into Mom. Sort of. It was surprising because brilliant, always-curious Madeline passed into eternity in 2005. When Mom passed on, her body was cremated, and her cremains placed in the family cemetery plot, near those of my dad, and my sister. Cremains, also called ashes, are dense and heavy: the elemental, net sum of our physical selves, and a box so compact, so small.

The funeral director also put a tiny portion of Madeline's remaining gray, grainy minerals into a pocket-sized brass urn with an ornate lid. We spread a bit of Mom's cremains from the small urn at our family property in northwestern Wisconsin, not far from Superior, the city where Madeline was born, grew up through much of her childhood, and earned her university B.A. degree.

Then we stored the little urn somewhere safe in our house. And lost track of it. As I sorted through items in one of our cabinets the other day, I found it...and accidentally bumped it over. A few grains of Madeline's minerals fell out onto the wood shelf. That got me thinking.

Minerals and water are what we consume to live. They are the essentials from which we are made. Minerals are what is left when the water and the life go away.

Now, hang on tight. Here is a major cognitive leap...

At our little primitive farm, we raise our small fiber flock of registered Icelandic sheep for their wool. Each day, we feed our sheep a measured ration of required supplemental minerals. These minerals are not found in their hay nor in our local soils and hay, and are tiny, angular gray and brown grains composed of selenium, cobalt, and other essential elements, born in the fusion reactions of distant stars.

Our own bodies are made of the same things: iron, calcium, sodium, potassium, and the numerous basic ingredients of our physical beings; all of these elements formed billions of years ago, then spread throughout the universe, eventually coalescing and forming into clusters, stars and planets and more. We are made of elements from our own planet and from our local star, the Sun, probably from others as well.

Everyone I love, everyone I treasure, everyone who tries my patience...we and all other living creatures are linked. We are made of the same minerals, with water and *life* added. Remove the water, remove the life, and we have only minerals. Dust to dust. Ashes to ashes. Incredibly, that amazing life force which somehow animates the minerals of all living things also unifies, humbles and mystifies all life. We each seek to know and honor that animating force in our own ways.

Take away that invisible power, take away the water, and we become reduced to basic materials, desiccated to component elements. This *wonder-full* animation we call life, that power causing us to move and feel and create and to re-create, is what remains as our universal mystery.

Fragile, that's what we are, easily deconstructed. We are all simple, yet so complex and fragile. Because of that, let's be kind to each other and to all forms and occurrences of life.

Throughout each of our active life adventures of many trips around the Sun, then back to our original, elemental state someday, kindness should be our mission. We are elements in motion, born from ancient stars, and deserving to give, receive and exchange respect with all other life forms, types and species. We all share these same amazing origins.

Thanks again, Mom, for yet another powerful life lesson after all these years. I'll love you forever. You have always been my Great Teacher. Once again, in a completely unexpected linkage of diverse ideas and concepts, you have reminded me of, and instructed me in, what is important in all our days of this life on Earth, our precious days so fine, so finite. And so Elemental.

Chapter 43

Hot Oatmeal, Albedo, And Al Gore

According to news reports several years ago, a senator in Washington, DC was prompted by huge late-winter snowstorms there to build an igloo-shaped snow shelter on government grounds. He put a sign on it saying "Al Gore's New Home" as a satirical political dig (pun intended) at the former vice president regarding his stance on global warming. The cunning senator was attempting to convey something like "Global warming? Right! I'll show you two feet of global warming on the White House lawn."

His understanding could be an easy conceptual leap to make, and it may be persuasive to many doubters of global climate change. Here are my own thoughts on the topic, with nuance and admitted bias.

Snow is a complex, wonderful substance. I know snow well, having moved and skied in so many sleek, efficient ways *on* it, slept *in* it, shoveled it, driven in it, and studied its physics in both academic and folk contexts. Simply put, I am a winter lover: an engaged, indeed a true winter-obsessed inhabitant of Earth's

Northern hemisphere. Snow is my friend, my passion. I love it. I know it.

And I know that two or more feet of snow at once in a southern area of our continent where snow is seldom seen means our global pot of variable weather is being stirred by heating. Our home kitchen helps provide understanding of what is happening.

As I watched our morning oatmeal boil today on the stove top, I realized that it represents a simplified version what is happening on our Earth: atmosphere and climate are warming, just as my water and oats were. And, as warming continues in the stainless-steel pan, causing the oatmeal to swirl and move within the water in three dimensions, so, too, does a warming planetary climate make moisture and warm-air/cold-air boundaries mix. The types of snows seen in the eastern areas of the US lately come from air temperatures of about 34 degrees Fahrenheit, so snow is not a harbinger or diagnostic sign of cold. It is just the opposite. Snows of these types and depths indicate a mixing, a stirring of the atmosphere due to heat disturbance, just like my oatmeal swirls and gyrates when I turn up the heat on the stove burner.

This brings us to albedo. *Albedo*. Not libido, although heating and stirrings could be involved in both. Albedo means the reflectivity of a surface. A shiny or light-colored/white surface has a higher albedo than a dull, rough surface. Think of your waxed tile floor versus your dark carpet. The tile is shiny and brighter, and it has a higher albedo.

So, here is the catch: warming and mixing of our planet's climate over the long term leads to more mixing up of warm, humid air with cold air, and to more snow over more surface area of Earth in areas that don't usually get snow. The more snow we have, the higher the albedo of the planet as it warms. At some point, the earth could be reflecting so much heat back into space that global climate suddenly goes the other way and snaps rapidly back into global cooling.

A new ice age? Snowball Earth? It has happened before.

Our senator friend in DC may be on the right track. Al Gore and all of us in this hemisphere and on this planet are in a period of massive global climate change. Perhaps a warm, cozy snow cave could be in our future. Hmm. I have always slept so well in snug snow caves.

By the way, isn't it interesting, from an always-curious, winter-nerd writer's point of view, that:

1. Ice ages on Earth seem to happen about every 20,000 years.
2. Several thousand years have elapsed since we last had a major ice age, with the Wisconsin period of North American glaciation ending about twelve thousand years ago.
3. Most of the scholarly, religious and historic writings we now base our civilizations on are from warm-Earth periods of not more than the last five to seven thousand years.
4. What would those people who lived through the last ice ages have written, if they could, that might change our ways of current thinking?
5. What will we write if we are the now-literate ones, with ways of preserving our writings, who may go into and live through the next world cooling and major ice age?
6. How might future world religions, science, art, culture, commerce, and all of life be changed?
7. And...What if we and Al Gore are all someday writing from our warm, protective snow-cave homes? Will we share our oatmeal with each other? Will he share his with us?

Chapter 44

Engineering For The Right Thumb

When your wife is an accomplished fiber artist, you can only watch and marvel. Debbi's abilities to envision, then to fully engineer functional garments and wearable art from fibers and fabrics of all types, are amazing. Her skills are broad and deep, as evidenced by her complex hand-knit sweaters made from our own flock's wool. From other types of cloth, she makes custom draperies, bridal gowns, sport coats, shirts, packs, purses, dresses, blouses, and ingenious adapted clothing for humans and animals with special needs. Her works are always stunning, clever, and expertly crafted. Debbi's creative fiber visions and skills developed early in her life. Her father was an early recipient of her gifts.

Debbi's father, my father-in-law, John, was a remarkably resilient, intelligent man. As a precision craftsman and dairy farmer, he was also a talented welder: a metal artist, who farmed so he could weld. John created inspired metal solutions for broken parts of important farm machinery and vehicles essential to his neighbors and himself, long before ready delivery of replacement parts to rural areas existed.

In the hilly Wisconsin farming community, where John took over the family farm when his father died in 1942, my father-in-law became the resident welding expert. Every farmer often needs metal parts mended and modified, and even fabricated from scratch when replacements are not readily available. When critical components of a baler, tractor, wagon or combine break during demanding haying or harvest seasons, farmers know the pressures of getting that important fix done quickly and well. They must stay ahead of the weather. John welded and formed his custom solutions during growing seasons, but he also had to make and modify metal parts in winter, often inside his wood-heated shop, but sometimes outside in extremely cold weather.

As a right-handed welder, John would take off his right glove to work barehanded. Small parts such as rivets, bolts, screws, chain links and pins are hard to manipulate in severe cold with gloves on. Thumb and forefinger prehensile abilities in the welder's dominant hand are critical to safe, accurate, and in John's case, elegant repairs. His right thumb constantly endured the penetrating nip of raw, cold metal in winter.

Debbi, age eight, saw her Daddy's plight each winter day. His right thumb was often sore and stiff, and she became determined to remedy his black, frost-bitten winter thumb. Debbi couldn't weld, but, as a budding fiber-artist nerd, she could knit, using skills and design wisdom she had learned since age four from Johanna, her Norwegian great aunt and a master knitter.

Debbi thought it all through, experimented with needles and yarn, then visualized and knit her own design of a protective *thumb* mitten. It even had a special crocheted wrist strap to secure it to her Daddy's frost-injured hand. She engineered it as the right mitten for the right thumb. Debbi's perceptive winter gift of childhood creativity, plus her skill, insight and love, warmed John's hand and his heart in deep cold.

Chapter 45

Dad Said Zed. Mom Said Zee

ABCD EFG HIJK LMNOP QRS TUV WXY ZED. Yup, ZED.

My Dad, Victor, had a French-speaking French-Canadian immigrant father and a German-speaking German-immigrant mother, both of whom said the final letter of the American English Alphabet as "Zed."

My Mom, Madeline, was the oldest child of Norwegian and English immigrants to the US and Canada, and she usually said "Zee." But sometimes for her it was also *Zed,* especially as she practiced her Norwegian language skills. Those European and Germanic language roots were strong in both of my parents' linguistic backgrounds.

Victor was a pharmaceutical chemist, and always had interests in science and international radio. He shared them with me. In his profession, and in shortwave radio, the last letter is always said as *Zed.* It still is in international pronunciations, in order to distinguish it from the spoken letters B, C, D, E, G, P, T, and V, and numeral 3.

By the way, International Morse Code for Zed or Z or Zulu is dah-dah-di-dit. Thinking of it as *"Dadda Did It."* makes a convenient way for me to remember the code and my dad.

Mom's *Zee* and Dad's *Zed* taught me there are differences in the world, and that different points of view, different approaches can work out well. For them, that included 65 years of marriage, shared work, and much fun with learning and acquiring skills and knowledge all their long, studious lives.

When we have a chance, remind us to discuss the myriad other differences that people can have...and still get along with each other.

We will list them all. From A to Zed.

THE TOPIC of Zed versus Zee has been on other writers' minds. Author Harrison Panabaker's excellent article in *The Canadian Encyclopedia* adds considerable detail and clarification. He cites the many historic names for the letter Z: *zad, zard, ezed, ezod, izod, izzard and uzzard*. Panabaker also notes the fifteenth-century Middle English *Early Dictionaries* brief mention: *"zed, which is the laste letter of the a-b-c."* Moreover, he quotes the interesting Z entry in Samuel Johnson's 1775 work, *A Dictionary of the English Language*, wherein Johnson writes of *"zed, more commonly izzard or uzzard."*

This quick summary above demonstrates that we curious word folks have much more to read and learn. The story of Z is more complex and fascinating than we might have anticipated.

Take your choice of zee, zed, izzard, uzzard, dah-dah-di-dit....or any of the other options we may use to convey that same phoneme, the speech sound /Z/ (zzzzzz). All of them work, depending on our situation and time in history. They represent the same phonemic concept, if you want them to. Variety in speech and language and life can all be wonderful, adding nuance, depth and joy to our words.

Chapter 46

Isle Of Pines Or Pisle Of Ines

Silly word humor. Can't beat it. And *metathesis*, that quirky, comical topic in applied linguistics, supplies some of the best laughs. Metathesis is an unusual, big term for a common thing we hear and do in speech.

Here is a plausible sentence anyone might say: "It is a beautiful day here in the country. Time to take a walk and listen to the birds chirping." Now, here is that same sentence with metathesis applied: "*It is a deautiful bay cere in the huntry. Time to wake a talk and listen to the chirds birping.*"

It's my parents' fault. They infected me. I never let on. Word humor. Word reversals. Word tricks and substitutions. All of these were their private comedy shtick, and are frankly just plain funny. I got it, at an early age, laughing to myself each time they were excited about a word prank they thought I didn't hear. It is still within me after all these years, a lasting gift from Victor and Madeline. Thanks folks.

Both Maddy and Vic grew up close to the lakes and forests of northwestern Wisconsin, graduating from Rice Lake High School. A popular fixture of their local landscape, and a lasting draw of young-folks culture in the lake, itself, was a small adventure island near the west bank, on the town side.

The Isle of Pines is a block or so in size, and about a block's distance off shore, an easily-swimmable stretch of water, or a quick canoe or boat ride. It was also a main party spot for young folks of the 1920s in Rice Lake (Lice Rake), my parents said. They spoke of it in prohibition-era details that I didn't understand in my younger years, always saying it with first sounds reversed: *Pisle of Ines*, not Isle of Pines. Pisle of Ines was a center of high school debauchery in their youth. Maybe still is.

Years later, in my university linguistics courses, I learned about sound and syllable reversals, and how some can be humorous. So, yes, the official term for it is *metathesis*, wherein the speaker reverses first sounds of adjoining words. "Blow your nose/Know your blows." "Flying saucer/Sighing flosser." Now you will keep on thinking of many others. Sorry.

"Spoonerism" is another common name for this effect, assigned from the speech habits of legendary British scholar, author, and orator, Reverend Spooner. He is famously remembered for announcing to his Oxford congregation that the next hymn they will sing is "Kinkering Kongs" (Conquering Kings). He is said to have introduced visiting royalty by asking his audience to please welcome "Our Queer Old Dean" when he meant to say Dear Old Queen. Reverend Spooner spoke angrily another time in public to alleged student arsonists who set ablaze a university commons area, accusing them of "fighting a liar" in the quadrangle. In a chapel service, he also reputedly stated "The Lord is a shoving leopard." and asked, "Is it kisstumary to cuss the bride?" at a wedding rehearsal.

Intentional and unintentional Spoonerisms are fun parts of word play and humor in English to this day. I admit culpability, learned from my dorky parents who made these up all the time.

Curiously, Spooner, Wisconsin is just a few miles north of Lice Rake. Hmmm.

A major precipitating word-nerd event happened in my teen years, when I was in the community theater production of Thornton Wilder's classic play, *Our Town*. The play takes place in the small New Hampshire town of Crawford's Notch, which, of course, became spoken by our cast, after about the second minute of our first rehearsal, as: "Nawford's Crotch." We cast members were never safe or sane with words thereafter.

You get the idea. Mom, Dad and Metathesis. Spoonerisms if you prefer: all spontaneous, weird and funny. My folks are ultimately responsible for my reflexive, internally convulsing chortles that I lapse into before jolting myself out of it in otherwise serious conversations. I can't help it, and am grateful my word-wracked (wrecked?) brain seems to always go there. It's now automatic and subconscious.

So next time I try to wax humorous in front of people, drawing on my warped inspirations from Maddy, Vic and Reverend Spooner, I will just relax, improvise and --get ready-- verbally *hoot from the ship*.

Chapter 47

Calculus Of Rlu

"To urinate, is pretty great..." an NPR show host once announced. Another public radio program was on in the background at home while I was working. I was not paying close attention to the scientist-physician who was speaking. She talked of human biology and evolution, but didn't fully catch my attention until she said "...*most of human life can be understood through urology...*" That's an imperfect quote, I'm sure, but it captures her major point. Humans, along with all other mammals, owe a lot to urination, and to the surrounding practices, customs, anatomy, and physiology related to our urinary systems.

My thoughts immediately went to our first, friendly family dog when our boys were young. *She* always lifted her leg to urinate. *She* marked her territory wherever she was. *Shiku Ukiuk* shared nearly 14 years of adventures with our boys and us, as we all grew up together in the 1990s and early 2000s.

Shiku was an American Eskimo female; a pure-white, fluffy, feisty, puff ball--kind of a miniature Samoyed--and the nastiest animal to others. A female dog is correctly referred to as a "bitch." She embodied our colloquial understanding and usage of that term. Shiku Ukiuk, meaning "Ice Winter" in native Inuit language, was our snuggly friend and family companion on a thousand adventures throughout her life. And she almost always practiced RLU. We seldom saw her squat to pee.

RLU means "raised-leg urination." We see it often with male dogs, and observe that female canids also do it. Wolves, coyotes, and both genders of other candid species mark their world this way. They use the radius of the scent imprint from their foot pad to height and distance of their urinary scenting on a tree or post to convey identity, presence, and their status in the local dominance hierarchy. Wolf packs typically allow only the alpha leaders, male *and* female, to do RLU. All others in their pack of both genders, must squat to urinate, or they will be attacked and bitten, possibly killed. RLU is a privilege only for those at the top of the pack...or those trying to be, often at their own peril.

As we are now in yet another intensive and offensive election cycle, observing the social posturing of humans in politics, business, medicine, education, and too many wars, we realize that figurative and even literal RLU and its associated customs and burdens are not limited to just dogs and wolves. That NPR scientist was correct: Urology may provide the golden key to all other human understanding.

Elaborating on all the above, here is another go (sorry) at the topic of urination. Let's subtitle this part *Scent of Woman*.

A woman *shepherd*, that is.

My durable wife, Debbi produces her own complex, singular brand of biochemical predator repellent for our wild farm. We pour and spread it around at our sheep paddock and grazing area to signal to the wolves, bears and others in our neighborhood that Boss Ewe Debbi is around.

Don't mess with her or her sheep.

It works. The scent of my wife, Debbi, the good, actually the *great* shepherd, alerts predatory, carnivorous residents of the forest and pasture lands where we keep our sheep that this territory is taken, and that we respect their presence.

Debbi's fluid statement is "We will keep our flock here within our safe boundaries. You are welcome to pass by as you wish, just don't stop in here for any snacks."

We have all gotten along fine, with no problems over our eleven years, so far, in tending our durable, rugged wool flock of registered Icelandic sheep. Mutual respect works.

As I understand human law, most of it is about setting and adhering to boundaries. So, how simple it is to use the scent of woman, and occasionally of a man, to affirm those spatial and behavioral limitations with our wild neighbors. Our local wolves, coyotes, foxes, fishers, cougars and bears have been good, fellow wild inhabitants because we gently, but firmly, let them know we are present and watching.

We respect each other and go our own ways with no difficulties. How relaxed and easy it is for all of us. No problems. Our flock are assured by our constant scenting that they see and smell from their shepherds. Our firm, impenetrable Sunny Cove Farm fences, along with our shared definitive scents and rules, continue to make for good wild neighbors of all the many species with whom we share our integrated, diverse sheep-land forest and fields.

Chapter 48

Bear Too Hot

On a brilliant, sunny Thursday afternoon, Mimi, our two-year-old granddaughter, and I sat in sunny third-base side seats at Brewers Stadium. A hot, bright August day in Milwaukee was in progress. We enjoyed the game with our growing family, all of us cheering in the direct, intense mid-afternoon sunlight.

Mimi's little blue and white Brewers bear was getting his fuzzy ears chewed as she snuggled him. Yes *him*. Definitely a boy, she assured me. By midafternoon, the sun became even more intense on us, and on many of our thirty-eight thousand or so stadium-seat neighbors.

Soon I heard a little voice on my knee saying "Bear too hot. Bear too hot." It wasn't the bear.

It was Mimi. In perfect 26-month-old speech, she conveyed the critical message we were all sensing: *Too Hot!* She and bear and I stood up, walking quickly up the stadium stairs to the much-cooler, shaded inside mezzanine area. It was packed with many way-too-hot baseball followers gratefully gathered under huge overhead ventilator fans.

"Bear too hot. Bear too hot." Mimi continued her precise, concise message until we were situated in shade with some water to drink, enjoying moving air sweeping from the big ventilator blades.

So fellow word nerds, think about this: Mimi used a noun subject, *bear;* a comparator, *too;* and a descriptor, *hot*. Better than formal structure, her terse words were efficient, compelling utterances. They worked well. At age two, Mimi deployed her growing capabilities with holophrastic, *telegraphic* speech: compressed spoken messages which convey just the important stuff. She transmitted only the essential words, as we all might do in a telegram or in a text message.

Mimi used her developing mastery of three central components found in all language systems: form, content, and function. Form includes *phonology* (speech sounds), *syntax* (word order), *morphology* (word structure), and *prosody* (pacing and melody of speech). Content refers to which phonological sequences, *word meanings,* to use. Function is the *pragmatic utilization* of both form and content skills to get your message across in real communication settings with varying partners.

These three main components of language comprise our foundations for complex literacy across all listening, speaking, reading, writing, and thinking. When we write, we make marks on a page or screen to signal what we could do with our mouths if we were to speak what the marks tell us. We usually call that *reading*.

Form, content, and function...all are essential components underlying human language. And Mimi used them so easily. On a sunny, hot August day at a major-league baseball game, Mimi with her bear, along with her wonderful words, reminded me again of our shared miracle of language.

Chapter 49

Center Of The World

It's a big world. London and New York, financial centers. Paris, fine-art and fine-food center. Australia's Great Barrier Reef, surf center. Tokyo's Ginza, electronics center. Hollywood, film center. It goes on with more information and places than we can process in one reading.

We can each name places that are our own centers of the world, and all these choices, these options, can be *bewildering*, literally: they can make us long to *be wilder* in our imaginations and understanding.

There *is* a wilder, simpler-yet-grander perspective. Black Elk, prominent spiritual and social leader of the Oglala Lakota (Sioux) Nation, taught "Everywhere is the Center of the World." He passed along to us much from which we can learn. In his long life from 1863 to 1950, Black Elk lived in dual cultures, becoming a recognized leader of his Nation and in his church. He saw no contradiction of values between cultures, believing the Creator made all of the world, and that all places on this planet can be sacred and special. Acclaimed sites or cities are no more qualified as the center of this world than are any of the other simpler places on Earth, no matter how remote or mundane they seem.

Black Elk was right. Here in our ancient, deep pine forests, or on bright Lake Superior shores, we are neither closer to nor farther from our Creator, or from any center of our world, than we are in Santiago, Reykjavik, Manaus, or Perth, or at our forest and farm.

Anywhere, *everywhere* on this planet, we are at *the* center of the world, in our Creator's presence, equidistant, without preference. The Center of the World for each of us is exactly where we stand in awe and honor of our universe to pause, reflect and worship right now.

WE WHO LOVE the wild often find our life inspirations outside. Our fascinations are the tall-pine forests, and the deep-blue seduction of pristine, clear lakes. Through early morning cold and snow, on starlit frosty nights, golden autumn days, and in full summer sun and winds, we trek our Northland trails and backcountry ways, where we overlook Lake Superior. Our shared trail and farmstead adventures are precious and beautiful, each in ways always changing. Caring for our flock draws us outside in all weather, at all times and seasons, connecting us with deep forests and rolling pastures here in far northwestern Wisconsin.

We especially revere Lake Superior, just over the trees and down the hill from our little farm. Superior, the greatest freshwater lake, so large, so blue, so wild, is the panoramic world we view just north of our paddocks. What sights and sounds of true Northern boreal forests, and of our planet's broadest inland sea await each day. We are continually awed and humbled by the power and *wildness* of this world. We are respectful participants in our northern paradise each time we hike or ski hilltops overlooking Lake Superior. Our bright, shining northern sea beckons.

We are reminded of Thoreau's words "In wildness is the preservation of the world." His writings influenced the modern environmental movement, and echo sentiments expressed by our indigenous and First Nations leaders for centuries. Their teachings still guide our lives.

Be sure to note that Thoreau wrote *wildness*, not "wilderness", as we might be likely to say today. Perhaps he meant what we now understand as that latter term. Maybe he was getting at something else.

A measure of wildness, in all its meanings and nuances--a wildness of thought, spirit, and place--is part of and key to preservation of ourselves and our world. In other words, we each need to conduct our lives a bit be-wildered (*of the wild; acting as a wild one*) to creatively cope with and thrive in our challenging, self-domesticating human societies, as well as in our changing natural environments.

Directed, refined bewilderment, is at the heart of all true human passion and innovation. In both wildness and in wilderness can lie our personal and planetary preservation. We are grateful for our forests, Lake Superior, and our tough, tenacious Icelandic sheep, mentors all, for teaching us more each day in every challenging, humbling, sacred Northland year. In its most positive sense, a bit of bewilderment each day is essential to each of our lives throughout all of our years, as we each seek to find our Center of the World.

Chapter 50

A Sewing Awl

Erika, age seven, the youngest child of our close friends, greeted us at her front door. She and her family hugged us, then showed us around their ornate woodland home overlooking Lake Superior. Their panoramic, vibrant view of Earth's greatest lake inspires, and humbles.

I sat down in a comfortable armchair in their front room. Quiet and polite, Erika brought a small wooden box to me. Without speaking, she sat down on the footstool next to my chair, and reverently removed the box lid. Erika took out a compact raw-hide pouch that she and her family had made. She untied its closing laces.

Erika withdrew a pencil-sized, dark-green, solid metal bar, about one-quarter inch square on the larger end and tapering to a point on the other. It was made of native copper, hand-hammered and deeply weathered, covered with copper oxide: a sewing awl, once used for punching through hides and lacing pieces together with cedar roots or red osier dogwood strands to make clothing and footwear. With its forging marks and sharply defined edges still visible, the awl was strikingly well made. Its precision and evenness spoke of its maker's craftsmanship, skill, and desire for symmetrical perfection.

"*It's seven thousand years old,*" Erika whispered.

I was fascinated, and fully convinced of her claim because of Erika's demonstrable reverence for the awl. Her family gathered with us, discussing how she found it on private land in rugged northern back country during a family canoe trip. Erika spotted the awl, green and dark with age, lying amid rocks near the shore; the symmetrical form of a human-crafted tool resting in gray sands, between boulders in shallow water during a recent dry summer.

Someone had made it, carried it, used it. And someone had lost it there millennia ago.

Erika held the sewing awl in her small hand.

INDIGENOUS PEOPLES HAVE LIVED in this copper-rich western Lake Superior region for thousands of years. And still do. Recent studies indicate human activity here at least seven to ten millennia before today. Intelligent people have been finding, shaping and trading copper and other natural materials in this inland-sea area for a long time, living at the edges of glacial flow and subsidence.

So Erika was right. Our wide network of North American trails and waterways, with trade centering on copper 7,000 years ago and more, was real. It happened right here in northern Wisconsin, Minnesota, Michigan, and Ontario. In fact, one of the western Lake Superior trade routes of the Brule River to Upper St. Croix Lake to the St. Croix River to the Mississippi River is right outside our front window here at home. I am looking out through the mists of Upper St. Croix Lake as I write.

Copper Culture citizens recorded their history in toolmaking, art, and rock petroglyphs. Today, we mostly value written words, but there are other ways of documenting that you have adventured and persevered on this planet, in this land. Copper Culture peoples and their paleo predecessors lived, mined, transported, and thrived here long before us today. Their evidence, though not always well recognized by modern societies, is present and clear if we observe and attempt to understand.

In the greater perspective of years past, we, today, are much nearer in time to Plato, Plutarch and Pilate than Copper Culture societies are distant from them in the past. We are overlooking thousands of years of indigenous intelligence, language, culture, technology, religion and trade. Our shared, expansive human story is far wiser, older, and more nuanced than we currently accept.

Erika has made me think.

Chapter 51

Granny Knot Or Square Knot?

Knots involve fascinating physics. My wife and I are experienced shepherds. Over the years, we have found some individuals in our fiber flock of smart, always-canny registered Icelandic sheep have an unexpected grasp of applied physics. They can untie Square Knots.

Yes... *Really.*

We have seen it happen often. Their prehensile lips and sharp eyes give them unexpected skills. Numerous short ropes we use for securing temporary fence panels can become mysteriously and inconveniently untied. And we know who the perps are.

Flock members Haakon and Bjorne, both alert, cagey neutered rams, "mouth" the ends of our sweaty-hand-tasty Square-Knotted ropes on fence posts and pull hard. The knots quickly slip apart. Even though Square Knots are great in first aid and back packing, they're not so great when you need other secure fastenings. So, now we use Granny Knots.

Granny Knots?

As an Eagle Scout in a family of Eagle Scouts, I learned from my parents and youth leaders to *always* tie Square Knots. "The B*est* Knot...!" I was taught.

Later, in my intense rock climbing and National Ski Patrol years, I found that Square Knots *can* sometimes be dangerous. They may work well in many tasks, but can also pull out, untie, and let go of a load if you use them to join lines.

It turns out that Granny Knots, those dreaded errors of attempted Square-Knot tying for which Scouts and climbers often receive reprimand, can be more secure, often *safer* in some applications.

And our sheep can't unite them.

A well-tied Granny Knot will hold more reliably than a Square Knot in certain situations. The knot user must be alert and aware. Flexible thinking is good.

Square Knots --doing things the "right" way-- can sometimes get you into deep trouble. Knowing when to use a Granny Knot -- when to safely bend the usual rules-- may work better.

It could save your life.

So... Granny Knot or Square Knot? "Best way" or another more flexible, perhaps better way?

Knotty issues to ponder for our daily sheep-engineering decisions, and for all the other things of life.

Chapter 52

Cleopatra: Word Nerd

Anthropology can be especially meaningful when we explore that field of study as compressed within the life of one individual. Cleopatra (69 BC – 30 BC) is such a person. She holds a unique place in our conception of history, academic study, and gender expectations. Over the span of ages, we are closer in time to this brilliant pharaoh/queen, than we are to the indigenous peoples who thrived here in the Great Lakes region of North America more than 7,000 years ago. Viewed in that way, Cleopatra seems to be almost our contemporary.

Cleopatra, so often held as a model of sex appeal and feminine wiles, was a polyglot: a super linguist. Her exceptional linguistic abilities enabled her accomplishments as stateswoman, strategist, ruler and powerful communicator across diverse cultures. Facile with words throughout her active thirty-nine-year life, Cleopatra engaged in governance, science, health, plus personal and political logistics, via the power of knowledge she gained from languages she spoke, and books that she read and wrote.

Not just the beautiful seductress of legend, the singular queen is reputed by some modern sources to have had an IQ of 180.

Cleopatra spoke and understood at least five to nine languages, in which she conducted matters of state, business and foreign affairs, as well as her storied romantic and complex family lives. To do so, she would have needed linguistic foundations of both *performance* (sounding like a native speaker) and *competence* (understanding and combining complex wording like a native user).

As pharaoh/queen, she was a recognized, accomplished leader. Cleopatra's reservoir of superior abilities bolstered her powerful appeal, enhancing her radiated equality in interactions with forceful male rulers of many nations as she served as last pharaoh of Egypt.

Known to formal history as Cleopatra VII Philopator, this exceptional woman of the Ptolemaic Greek dynasty, was its only member to learn Egyptian, the language of the people she ruled. Although ancient sources report and legends hold that she was competent in many languages, current hard evidence offers no proof that she spoke anything but Greek.

Nonetheless, her near contemporary, Plutarch (AD 46--after AD 119) in *Parallel Lives*, 27, 3-4, wrote that Cleopatra switched in and out of languages including: Greek, Egyptian, Median, Aramaic, Arabic, Hebrew, Troglodytic (Persian), and Ethiopian. With her linguistic gifts, she likely picked up Latin, too, during her interactions with educated, powerful Roman men and women as well as her household servants.

Plutarch's depiction of Cleopatra's language skills affirms the central premise that she could adapt, at will, to communicate. Cleopatra had intelligence, finesse and motivation enabling her to interact with others in their own words, spoken and written. History confirms her strong leadership.

Cleopatra also explored scientific method in her personal laboratory, experimenting and writing on cosmetology, health and herbal healing. Her formidable linguistic skills allowed her to read works of others, as she also *wrote* books on varied topics. Her books were lost in the Library of Alexandria fire, 391 AD, yet Cleopatra's influence on science and medicine extended into early centuries of Christianity. Her legacy was continued by her brilliant, devoted daughter, Cleopatra Selene, who honored and preserved her mother's exceptional life and memory.

Perhaps not as obvious to our present-day thinking, a significant, practical indication of Cleopatra's wisdom, abilities and functional impact on the multi-national ancient world is that she was not killed off early by powerful Roman men. Viewed as their competent, canny competitor in politics, governance and commerce, she survived and thrived. And here we are, still writing and thinking about her today.

Cleopatra's historical longevity is often attributed to her physical beauty and sensuality, yet her powers were more formidable than that. Her armor of accomplishment and aura of daunting, proven abilities surrounded the amazing pharaoh/queen with an intimidating cloak of protection. Cleopatra's stellar life and influence, even into current and future times, demonstrate the benefits of her linguistic talents and intelligence as bases for achievements to be remembered for the ages.

Chapter 53

Children Of A Lesser Doc
-revisited-

Sitting in packed rows at a medical school graduation, I overheard this conversation behind me as the Doctor of Philosophy (Ph.D.) degrees in medical sciences were awarded. "When will they get to the real doctors?" The reply was memorable. "These are the real doctors. They discover, verify and teach knowledge that allows office doctors to do what they do." How true. Doctors of Science and data are entrusted to find, test and teach discoveries, passing forward information and skills that enable progress in medicine and in all other modern fields.

Doctoral degrees of varying types in academics, theology and health fields signify intelligence, persistence and achievement of the holders. In the early fourteenth century, our English use of the term "doctor" began to switch from academics and doctors of the church (learned teachers of doctrine) to medical care providers. Now, doctors' degrees for health-care providers and other "Docs" are many and various, with pecking orders in the "What-kind-of-doctor-are-you?" food chain.

My sons and I, it seems, are all children of a Lesser Doc.

To vastly oversimplify, my premise is that holders of the MD,

Medical Doctor, the most familiar of US medical degrees, currently top that Doc food chain. Neurosurgeons, cardiac and orthopedic surgeons likely reside at the pinnacle of that Top-Doc pecking order. Within this descending MD lineage are other surgical specialists of all types, then general surgeons, and then other MDs who practice medicine but not surgery. Internists, psychiatrists, hematologists, among others, may be in that group. US-based culture and media often place the general practice, pediatric and family practice MD physician near the bottom of this top section of the Doc hierarchy. Other nations and cultures view and value medical practitioners' degrees and roles differently.

The next grouping of clinical Docs in US media, I submit, are holders of the DO, Doctor of Osteopathy, a degree with less popularization and thus less public familiarity. Doctor of Osteopathy (DO) is a valuable, common but still not well-known medical degree in the US. Do you know of a TV show titled "Emergency DO?"

Nonetheless, DO physicians of all specializations *are real doctors*. Yes, they offer a verified, different perspective: an osteopathic view of medicine instead of the allopathic bent of the MD world. DO-variety Docs practice in usual physician roles and specialties to which we are all accustomed, depending upon their training and licensure, from neurosurgeon to family practice. But somehow DO Docs still do not get the wider respect and acclaim that MD Docs do.

My grandfather, Earnest, was a rural, family-practice physician, also a prolific artist, author, woodsman and a learned lover of the wild. He retired to his Northland log cabin on a lake, and for a time, owned and cared for this cherished, old-growth family forestland. DO Grandpa Earnest, always called "Doc," was, in the view of some, a Lesser Doc.

Following the Docs of MD and DO kinds, in order, could be the other surgical and invasive-procedure clinician Docs who poke, prod, cut, drill, and stitch us and critters, as well as treat using medications in their practices. These include oral surgeons, dentists, podiatrists, veterinarians, and other essential healers. I would also list chiropractors, optometrists, and audiologists in this group of doctors who test, touch, treat, medicate, operate, rehabilitate and/or prescribe for us and our family members of all species. They hold and practice with great skills, knowledge, professionalism and merit. Yet pop culture often says these healing professionals are not "real" doctors.

Following this regression of health care providers toward the even Lesser Docs, may be those Docs who, according to some wisdom, "can't really do you any good." These could be Doctor of Philosophy, Doctor of Education, Doctor of Science, Doctor of Psychology degree holders, and similarly degreed healthcare and science professionals across a variety of "softer" fields, typically areas of highly specialized though largely non-invasive clinical practice. Although these Docs are reliably and rigorously well-educated products of grueling, difficult academic and clinical curricula, internships, residencies and post-doctoral training, they are often regarded as peripheral, even comedic.

So, our sons and I are, indeed, all children or (great) grandchildren of a Lesser Doc: an Ed.D. father, and a DO (great) grandfather. Somehow, we seem to have managed, even in our lowly Doc status.

Curious, isn't it, that the term "doctor" derives from the Latin word, *docere,* meaning *to teach,* and that the "physician-as-doctor" concept developed from the learned academic degrees of the medieval world. Doctors of the church and of academia were wise, disciplined leaders. Medical professionals adopted that noble, advantageous title to add respect to their then often fumbling roles. It is interesting that surgeons and dentists were once considered mere technicians, indeed barbers and bleeders in the latter case, not doctors. University medical education for US physicians became mandatory only late in the 19th century, existing as primarily an apprenticeship field prior to that pivotal time of increased call for rigor and ethical, efficacious practice.

It makes you wonder who the real doctors are. If educating diverse patients and learners to create curative change is at the root meaning of being a "Doctor," then how many detached, uncaring physicians and other clinicians and scholars who bear that venerable title may actually be the true Lesser Docs?

Chapter 54

Winter Love

Winter is great, especially here in the Lake Superior Northland. Why? Two words: No Bugs. Winter is also great because our homestead barn turns into a huge walk-in cooler, so handy for all those special holiday and weekend leftovers and beverages, often from Halloween way through Easter. What fun to just walk in and grab from an assortment of cold treats off the shelves.

Winter also adds constantly evolving three-dimensional sculpture to our rocks and logs around our Red Pump, throughout the forests and across the lakes. Our huge glacial erratic boulders scattered around our deep Norway pine woods become artists' dreams, always changing in light and form with the drifting, blowing snow.

Another huge reason I love winter is that ice and snow can make us, for a few months of the year at least, almost superhuman. We can slide faster, fly farther, and float smoother on our skis, skates, and sleds, under our own power and with just a little help from our friend gravity, than by self-powered methods at any other time of the year. Ice and snow give us each capabilities to move in ways we simply can't in other seasons. During winters here, we all frequently, literally *walk on water*: the solid kind called lake ice. We can gain mechanical advantages of speed and grace, and a boost to our self-esteem.

Also, as with many, I love to ski. Any type of skiing. I've tried them all, and this is my 70th year as a skier. My folks started me as a ski jumper when I was two, then I switched to downhill skiing when I was about age eight. Since then, I have raced a lot, been a Senior National Ski Patrolman, Patrol Director, Ski Instructor, and have shredded on snow boards. Plus, I've done much cross-country skiing and skijoring over the years. Now, we often ski in our forest and on our lake from just out our front door, here at home in our paradise of rural northwestern Wisconsin.

The primary reason I love winter, however, has become more evident to me as I see our grown sons living actively and dealing with life's challenges. We taught them early on to ski and to race well. In slalom racing, if you fall, go off the racecourse, or "miss a gate", meaning you make a mistake and don't go through the pairs of red and blue poles according to race rules, you have two options: you can keep on skiing, hoping no race official notices and you don't DQ; that is, disqualify. Or you can stop, taking the time, effort, and embarrassment to climb back up the slope, and ski down through the gate the full, correct way. You lose the race, but you do what is right. When my wife and I coached our boys in racing, we always stressed how important it was to be persistent, honest, and to finish the full slalom course according to the rules, not just being concerned about finish times.

They learned well. As I watch them now in their successful adult lives, dealing with personal and professional challenges and frequently "climbing back" through the many metaphorical missed gates we all have in life, they have become *our* mentors. Life lessons they learned early in the cold, and that they live out well, are now the main reasons for my lifelong winter love. .

Chapter 55

Conspire

How odd can daily life get? I am writing on December 16, 2020, amid our worsening COVID-19 global pandemic. For us living through this viral invasion, our simple, common human act of conspiring, that is *breathing together*, meant in its best way of meeting and talking with others, has become dangerous. Lethal, viral RNA bits, aerosolized by each of us while speaking, singing and plain old breathing, spread unseen, unnoticed into the atmosphere around us in close quarters. This invisible monster has spawned *conspiracy* thinking beyond imagination. Yes, there it is: a transmogrification of that word-of-the-moment, *conspire*. It deserves consideration and discussion.

Conspire, in Standard American English (SAE), derives from its past-participle stem in Latin, *conspirare*. Conspire literally means "to breathe together." It has shared roots in both Latin and old French.

As we work, teach, learn, buy, sell and socialize, we conspire. We breathe together. Conspiring, in its truest, benign sense, is what we miss in these days of social distancing, our new term which really means "physical separation." We are not now sharing atmosphere, not swapping air with those around us, and that's tough to do. It means minimizing, even giving up family, church, school, theater, sports, and nearly everything else we modern hominins prefer to do with others in groups, with all of us breathing atmosphere together in shared spaces.

Breathing, or more properly respiration (*re spir a tion*: breathing again and again), is intimately connected with speech and song, both behaviors essential to human interactions. During rest, or *tidal* breathing, our in-and-out airflow takes about the same time for each. In speech breathing, we inhale for one unit of time, and exhale with force for about three matching time units, regulating exhalation to finish spoken sounds, both voiced and unvoiced, for syllable pulsing and word phrasing.

During song breathing, our one-unit inhalation is extended to five or more units of expiration, meaning exhalation of air to hold out sounds. Indeed, singing versus speech is characterized by changes in vocal frequency, pitch, and extended periods of forceful exhalation to sustain vowels, the main distinguishing element between speech and song. Consonants are more difficult to sustain in speech and song.

In SAE, we seldom use *inhalation phonation*, speaking on *inhaled* air. Our interjection, "Inquisitive *Huh*" (I wonder?), is made by breathing out. Companion, "Surprised *Huh*" (I'm startled!) is generated on a quick in-breath: inhalation phonation. Say *Huh* like you are wondering about the beep in the other room. Then say *Huh* like you are surprised. In the first one, air flows out. In the second, you pull air in.

From our early linguistic developmental days of infant crying and vocalizations, we become skilled with *expiration*, our outflowing breath: e*x spir a tion*. In some languages, infants hear adult speech models of breath going both out *and in*, as little ones incorporate those opposite breathing skills into future speech repertoires. Not so much in SAE.

Breathing is essential, so conspire, inspire, expire, and combine these as best you can for life. Someday, with other people, we *will* swap air again as we talk, laugh, shout and sing together. With science leading our informed behaviors, not lies and myths, we will make this happen.

Chapter 56

Eternity With Pinewood Mercury?

Some songs get stuck in our heads. Auditory memory and earworms can be curious and mysterious perceptual events. And could they become part of my possible eternity with Pinewood Mercury? Or, at least, will the constant replay of their interminable advertising jingle rattle forever in my internal audio player, making it seem like eternity?

Could be. Here I am, just mowing the lawn, lulled by the low buzz of the motor. My mind relaxes and wanders. Then, the jingle creeps into my brain. First, it is quiet. Soon, it emerges with full, raucous intensity and splendor of the 1970s AM radio commercial it once was:

> *It only takes a minute for you to take the wheel.*
> *At Pinewood Mercury you get a better deal!*
> *Something something something something*
> *Something something something something*
> *Save a dollar, save a day,*
> *Something something something something,*
> *Find your car...and drive away!*

Eternity With Pinewood Mercury?

*It only takes a minute for you to take the wheel.
At Pinewood Mercury you...*

Aaaak! Make it stop. I keep mowing. My feeble brain tries to fill in the missing parts, constantly looping back through the catchy phrases I can still remember, then skipping over the *something-something* parts.

Irritating! I try to shift my thoughts to something else to interfere. What awesome clouds, the sky ...*takes a minute for you to...* Akkkk!

OK. Forget that. Let's look at those comical squirrels. Beautiful sun through the trees. Grass so lush. Smells great as I cut it, making me sneeze. Wonder if I put in enough gas? Better push mower back to the shed and find out. Probably needs a fill. Yup, looks almost empty.

*... take the wheel.
At Pinewood Mercury you get a better deal.
Something something something...*

No. Pleeease...

*Save a dollar, save a day
Something something... Find your car.... and drive away
It only takes a minute....*

Sigh... It keeps on going. Can't take much more of this. Looking up on my phone. Hmm. Pinewood Mercury on Central Boulevard went out of business *thirty-six years ago.*

Good. And I can't find their song now.

Perfect. They're gone. It'll take more than a minute to get me to ever take the frigging wheel. Find my car and drive away. Hah! No more, thank goodness. Done. Go away. My circuits are overloading with that pernicious jingle. Worked for them. Not me.

Tank is filled. Caps are tight on mower and gas can. Pull the starter. Back to work. Purring motor is soothing. Look at all those birds at the feeders, trees in dappled sunlight. What a Great day....

... for you to take the wheel. At Pinewood Mercury...
Something something something...

NO... Make it stop! Eternity with Pinewood Mercury?

Chapter 57

Ice Mirror

From our garage hockey equipment box, I grabbed my Bauer skates, hockey gloves, and one of my old banged-up, ragged-taped sticks. Hadn't used them in a year or more, and it sure felt good to be heading for a well-lit skate on the glassy lake. With the total lack of snow in our area that year, the walk down to the lake bank was quick. There, I sat on a big rock at the edge, tied on my skates, and put on my gloves. I grabbed my stick, pushed off with my life-long skater's legs, and power glided out on my smooth, personally reserved rink. There was no one else around. It felt sublime to stride strongly, and skim over the slick, bright surface in the sun. I skated north about two miles in just a few minutes, where the ice got rougher, then I turned around and headed back south to explore around the Island at the other end of the lake.

In my rush to get out on the ice during sunlight, I forgot to bring a puck along to stick handle and pass as I skated up and down the lake. Our puck bag, of many black, hard rubber discs and a few red-plastic street hockey balls, was underneath other gear in our box. I neglected to grab it when I left. Not a problem for us old-time outdoor hockey guys; I just skated around any loose chunk of ice, small stick, or pinecone I saw on the ice, scooped it in with my curved stick, and passed it on ahead to my imaginary teammates as I skated hard. What fun, and what pure joy, to move nearly silently and effortlessly on a fast, friction free surface. "Best skate ever," I thought, even though I was alone on the lake. I skated along briskly, enjoying the wind from my speed, the endless blue sky, and the open panorama of the lake and surrounding forests. I was lost in my own exceptional early winter world.

Hey, up ahead, a dot on the ice. I skated up closer and could see it was a faded yellow round thing. A tennis ball? Or had someone lost their street hockey ball or their driveway puck out here earlier? Curious, I charged up closer, never guessing from a distance what it really was: a whole lemon, frozen solid. How random: a loose lemon lying on the ice, hard as a rock, and faded from a bright grocery store yellow. The story behind that was anyone's guess. Maybe it fell out of an ice fisherman's cooler? Maybe animals got it out of garbage near a lakeshore cabin?

Who knows? Whatever its method of arrival on the lake, the lemon was exactly the missing "puck" I needed. I skated fast now, in rapid bursts, and slapped it on ahead on the smooth ice, then raced to catch up to it. I stick handled it forward and backward, in tight turns and long loops; I tried crazy behind-the-back and between-the-skates passes.

They all worked; I was so loose and relaxed. My team of one was looking good. If only I'd had a partner to pass to or been in a game that day, could I have shown them my old stuff. I didn't even have to worry about hitting ruts or holes during all my gyrations, because the surface was so clean and clear, almost like newly re-iced indoor skating.

In a while, I was tired from all my tight turns, stops, and reversals of directions with my new lemon puck. It was my first skate of the winter, so to slow it down a bit, I skated directly north, around the east side of Crownhart Island, then turned south at the Island's northernmost tip to skate along the sunny west shoreline in the windbreak of the tall pines on the Island. I cruised in a silent, pure world now, sending my ice lemon way out ahead of me, then chasing it, and shooting it out again, over and over.

My attention had been fully occupied for the last hour, up and down the lake, back and forth. Now, suddenly, a moving shape came into peripheral view over my left shoulder, something moving casually, at about shoulder height and maybe just ten feet now to my left. It was a large bald eagle coming slowly to my level, in full-spread glide pattern with wings at least seven feet across, tip to tip. He, maybe she, floated in silently, about five feet off the ice, focused on the lemon out ahead of us about fifty feet. The eagle held his glide pattern over the lemon, dropping to about three feet above the ice, then tipped his head down to look more closely at the yellow treasure as he passed above. He gritted through a tight left banking power turn, not unlike what I had been doing on the ice surface. Then he rose in the air, flew back around me, and came in again from north to south over the ice at about waist height, trying to get a closer look at this mysterious object.

I stopped and watched in amazement. My silent lake partner repeated his close airborne inspections of the random thing on the ice at least three more times, trying to figure out what that wobbling, flopping, skidding, yellow object was: Food, fish, unusual prey? I can only wonder. Then he flew off, disappearing into the endless blue over the distant forests.

For a few moments, the Eagle and I were in a sunlit, clear world of curiosity that day on the ice. We shared our joy of moving freely and effortlessly: him reveling in gliding, swooping, and turning in the air; me enjoying much the same on my glassy two-dimensional medium of polished, once-in-a-lifetime lake ice.

We both pursued a silly target that afternoon, and we had some fun together. I will never forget us both chasing that random lemon during on our shared St. Croix ice mirror adventure. What an experience, and what a story. I could not make it up if I tried. And, by the way, our ice lemon is still in a plastic bag in our barn freezer near the Red Pump to this day. It will stay there as a reminder of that cold afternoon of crystalline perfection, frozen fondly in my memory, forever.

Chapter 58

Y Not Ur?

Errors in spelling and usage of *your* and *you're* are common. They could be minimized, even eliminated, at least in our informal writing, if we used the time-tested, 150-year-old telegraphic, now texting, simplification: UR (di-di-dah di-dah-dit). Let's just type UR, and let it cover all variations. It sure could be simpler and save a lot of time, agony and frustration with proofreading, editing, corrections, and many fourth -grade spelling issues.

Related data on word usage are interesting. The word "you" ranks at number 14 out of over 5,000 of the most common Standard American English words used in print. The possessive form "your" comes in at number 69, "yourself" at 919, and "yours" at 3,183. The verb "be" and its various forms, rank number 2 in common usage of printed words. Clearly, UR possibilities occur often in American English writing.

Texting and typing, as practiced nowadays, have derived from telegraphy protocols from a century and more ago to allow keystroke savings by means of widely accepted abbreviations. Our reliance on these proven rate enhancers emanate from 180-year-old communication shortcuts such as LB for pound, NO for Number (not for North?), SOS for Emergency, PKG for package, AM/PM for time changes, plus many others. Be sure to check these out in the *Phillips Book of Telegraphic Codes*, written for telegraphers in the late 1800s. It established, codified and promulgated many of the common journalistic and telegraphic abbreviations we still rely on today. These keying and typing helpers assist us in saving repetitive movements and keystrokes that are required of telegraphers and typists. They offer shortcuts enhancing movement of vast amounts of information and data by landline, radiotelegraphy, digital transmission and reception of data. (And we won't even mention confusions with the word "yore" which can sound so similar, but pertains to times in the past, not to the pronoun "you".)

For actual data on exertion and input efficiency, let's look at the practical numbers in this possible solution of using UR for *your* and *you're*. In addition to not having to decide which combination and punctuation to use for the letters *y o u r e*, plus, maybe, the apostrophe, a considerable cognitive time saving, using UR for each or either would result in saving two keystrokes for *your* and three keystrokes for *you're* each time they are typed. Those are common occurrences in written Standard American English, as shown in the above frequency-of-use data. This could result in a fifty percent saving of effort for the first (your), a sixty percent saving for the second (you're).

Throughout the entirety of our writing, if we estimate that one of those forms of *you* occurs at least five times in a page of 300 words, our practical result of savings in keystrokes, as well as grammatical or morphological decision making, could be substantial in our manuscripts, notes, letters, texts, advertising and other print applications, whether in handwriting or machine typing.

Electromagnetic telegraphy, per Samuel F. B. Morse and Alfred Vail, among others, originated and first became useful during the 1830s and 1840s. Since then, numerous sharp, canny telegraph operators, using only one or two switches --their Morse Code "keys" of various designs, to create electrical circuit openings and closures representing letters and numbers and punctuation-- have developed and figured out ways to make their movements as fast and effortless as possible. We carry this intent forward with our own daily digital-device use in texting and typing of abbreviations, expansions, predictions and macros.

Perhaps is it time, once again, to simply standardize and convert *your* and *you're* into their simplified, proven two-letter form already used for decades: UR...or just ur in lowercase. This could work well now, at least for our informal writing, and could offer us a significant reduction of keystrokes and confusions. Both would be pleasant results.

So, what are ur thoughts on this? UR going to be among those who decide usage practices and style over this next century. UR thoughts and ideas will influence my and ur typing and writing for decades to come. Please raise ur digital device of choice high if ur ready to leap forward to ur and my telegraphic-shorthand past.

Chapter 59

Stereognosis

Your screwdriver just fell under the shelf you are working on, so you grope in your toolbox behind you to find that spare you keep handy. With your eyes on the loose-hanging shelf bracket, you fumble your hand among the tools. And there it is. Just the screwdriver you need.

Or maybe... It is allergy season, and you are coughing while you drive. With your vison kept safely on the road, you hurriedly feel around inside your purse or pack on the passenger seat, bumping into your rescue inhaler.

Or maybe... You are visiting dear Aunt Jenny for lunch, enjoying a bowl of her homemade turkey-dumpling soup. So good, but in that third spoonful, your tongue notices something out of place. Without saying anything, because you like Aunt Jenny and her cooking, your tongue feels around in your closed mouth to identify the offending shape: a small piece of turkey bone. You must not swallow it, so remaining calm, you wipe your mouth with your napkin, quickly transferring the bone shard, as you enjoy your soup, and your relaxed visit with wonderful, fun Aunt Jenny.

If those scenarios seem familiar to you, then you have relied on our least-respected sense: *Stereognosis*. Stereognosis is the functional combination of our sensory systems, including touch, smell, hearing, proprioception, taction, haptic skills, and plain *common sense*, to find, identify and manipulate things without looking at them. In a way, stereognosis allows us to "see" with other parts of our bodies, without using our eyes.

If you have inadvertently sat down on a hairbrush left on a chair, or stepped on a rock inside your shoe, you know how fast you identified and dealt with those surprises. Eyes were not necessary. You probably handled the matters just by touch, without even needing to look.

One of our most-important applications of stereognosis occurs in our mouths and throats. Oral stereognosis monitors how we process foods, liquids and medications of varying shapes, types and textures: those pesky turkey bones, olive pits and other hard, sharp inclusions we don't expect, even broken teeth. Oral stereognosis also serves us in managing saliva, a typically unnoticed function, unless you are experiencing conditions that affect how you handle saliva production and swallowing.

As we build up saliva in our mouths from chewing, breathing and talking, we need to swallow it at regular, unobtrusive intervals so it does not spill out or become aspirated during our inhalations. This is typically an unspoken basic of biological functioning we all depend on. But people of any age, gender or race, who are living with cerebral palsy, amyotrophic lateral sclerosis (ALS), or other neurological injuries or illnesses may have considerable, embarrassing difficulties keeping sufficient saliva in their mouth for eating and speech...yet not so much that it *overflows*. Strategic handkerchiefs, tissues, bibs, and medications can help, but disruptions in saliva management during play or meals, or in educational, social or work settings can have significant negative impacts on anyone experiencing the problem. Unless you have dealt with this condition, it is hard to imagine. Stereognosis helps us to cope.

For a simple experiment, next time you take vitamins, put two different shapes in your mouth. Oral stereognosis will guide you on what and where each is, and on which order to swallow them safely. Be careful. And be grateful for our least-respected sense: Stereognosis.

Chapter 60

Perseverance And Ingenuity

Landed. Safe! February 18, 2021. NASA's *Perseverance* rover, with on-board companion, a four pound, dual-blade helicopter, *Ingenuity*, landed on Mars: intact, communicating, and readying for missions. These are dreams for the ages of space-exploration professionals, and us amateur space geeks. We all waited months, years, for this day, holding our breath during the voyage, anticipating a secure descent and landing. On the total-coolness scale of one to ten, this is at least a thousand. Amazing accomplishments happened today. More await.

On a shelf near where I write, is my Gilbert reflecting telescope, and "Robert the Robot," my best-ever Christmas gift in 1958. Made of stiff, jointed gray and red plastic, Robert operates from a handle-crank you wind to make him move. As you steer him on his wheels, a small, internal record plays "I'm Robert the Robot, mechanical man. Drive me and steer me wherever you can!" He was my eight-year-old wish and delight, and still sort-of works, at the end of his three-foot-long, metal, mechanical drive cable, cranked by this now much-older kid.

Robert was my zenith of remote-control toys at the time he showed up. But, as with many, I dreamed of cars, trains, robots, planes and spaceships controlled by radio waves, at long distances. Maybe even across the country, or out to the Moon. Eventually, I got my Ham radio licenses so I could operate Radio Controlled air, water and land craft on frequencies not interfered with. RC was so cool. Our future.

On this day, the full reality of AI-guided robot explorer Perseverance and her sidekick, Ingenuity, flying to and landing on Mars, sending back visual and auditory data, then touring parts of Mars by *radio control* and on-board AI, thrills me. Each day of their working lifespans I will follow their progress, troubles and discoveries, and marvel at the new knowledge they will transmit back to NASA.

After their 293-million-mile journey to Mars, the focus today was EDL: Entry, Descent and Landing at Jezero Crater. Space miles, carefully monitored, would amount to nothing if EDL failed. Around the world, we amateurs at home, along with NASA engineers and scientists, watched in anxious awe as the "Seven Minutes of Terror" elapsed. With communications blocked by plasma generated from entry heat on the lander, so many things could go wrong. We wouldn't know for 11 minutes. A thruster misfire, software glitch, even a strong gust of Martian surface winds as the Sky Crane lowered its cargo could have ruined the mission in its final moments. Mechanical damage from a crash landing would have been disastrous, plus, with thruster fuel on board, a hard landing could result in both rover and helicopter being destroyed by fire, even in the low-oxygen Martian atmosphere.

None of that happened. Perseverance uploaded photographs within minutes, and will drill core samples of Mars rock for return to Earth in future missions. We again realize how space exploration is generational. Just as Robert the Robot sparked lifelong fascinations for me and others six decades ago, I know that space scientists and engineers, some just kids today, will pursue future missions, inspired by Perseverance and Ingenuity.

Chapter 61

When Nerds Post Noontime, Office Door Notices

Marketing Agency: Out to a Satisfying Lunch of Nourishing Goodness
Nuclear Physicists: Gone Fission
Professional Boxers: Out to Punch
Adams Family: Out to Lurch
NASA Engineers: Out to Launch
Composers: Rest. Hold. Rest. Allegro
Snack Wholesalers: Out to Munch
Science Writers: Ingesting Midday Comestibles in Absentia
Gas Station Clerks: Filling Our Tanks
NHL Defensemen: Out to Crunch
Lawyers: Lunch Now. Will Be Back Suin'!
Foodies: Out to Lunch. Back by 5. If not, out to Drinks, Dinner, and Dessert, too
Fortune Tellers: Out to Hunch
Karate Masters: Break
Hobbits: Out to Breakfast. Second Breakfast. Elevenses. Lunch. Nap

When Nerds Post Noontime, Office Door Notices

Godzilla: Powerline Time. Will Stomp Back Soon
Banana Growers: Out to Bunch
Metal Formers: Getting Bent
Teachers of Italian Language: Chow
Symphony Conductors: Bach Soon

Chapter 62
Entertaining Garbage

In our scenic little village, folks take a lot of pride
In our forests and our lakes, our blue skies where eagles glide.

Some choose to cart our own refuse. We haul it all in stride.
Sometimes our timing isn't right. Our garbage takes a ride.

Drop-off days are clearly posted. Village forms inform us well.
Times when we should haul our stuff are easy to foretell.

So...
My tale is of extended grunge. Yes, listen. You'll deride
This story of my addled acts when garbage takes a ride.

Cans and glass I sort out quick. I put cardboard piles aside.
Plastics packed, bags are bound; now dragged and stacked outside.

Boxes flattened; papers racked. Stuff on car top and inside.
Our Subaru is loaded full. Too much. Too long. Too wide.

Entertaining Garbage

We back out now. My view is blocked. I can barely see.
Leakage. Rattles. Aromas. Wow. A Driving Dumpster. We.

We reach the entrance. Dump is closed. Recycling locked
with key.
I hardly can accept the thought: it all goes home with me.

It aggravates my ego. It pesters at my pride,
That once again, to entertain it...
I took our garbage for a ride.

**Life in our small rural village has its advantages and challenges.
We try to maintain a prevailing nerdy sense of humor about all.**

Thanks for reading *Noon at the Nerd Table*. I hope you enjoyed it. Please let me know your comments and ideas.

T. W. King

twkingwrites@gmail.com

Sunny Cove Publishing
PO Box 98
Solon Springs, WI USA
54873-0098

Noon at the Nerd Table is available on Amazon.com Kindle Direct,
in stores, and via email or postal address above.

About the Author

T. W. King writes in far northwestern Wisconsin, not far from Lake Superior. Tom and Debbi have lived in many different cities and towns, but love their homestead in this old-growth Norway and White Pine Forest the most. They ski, hike, skate and run, reveling in all seasons, and tending their flock of rugged Icelandic wool sheep. Tom writes daily, often across several genres, and creates new tales, poems, raps, music and songs of the Northland, performing with Debbi and their creative Northland folk band. See HALLBJORN.com

www.ingramcontent.com/pod-product-compliance
Lightning Source LLC
Chambersburg PA
CBHW060832220526
45466CB00003B/1075